2008 北京奥运——
北京奥林匹克公园森林公园及中心区景观规划设计方案征集

International Competition for Landscaping of Forest Park and Central Zone in Olympic Green

北京市规划委员会
奥运森林公园建设管理委员会 编
北京市园林局
中国建筑工业出版社

Edit:
Beijing Municipal Planning Commission
Olympic Forest Park Constraction and Administrative Commission
Beijing Landscape Bureau
China Architecture & Building Press

中国建筑工业出版社
China Architecture & Building Press

图书在版编目(CIP)数据

2008北京奥运——北京奥林匹克公园森林公园及中心区景观规划设计方案征集/北京市规划委员会等编.—北京：中国建筑工业出版社,2004
ISBN 7-112-06298-5

Ⅰ.2… Ⅱ.北… Ⅲ.①夏季奥运会，第29届-体育建筑-设计方案-北京市②城市规划-设计方案-北京市
Ⅳ.TU245

中国版本图书馆CIP数据核字(2004)第001990号

责任编辑：郑淮兵 杜 洁
版式设计：付金红
责任校对：张 力 张树梅 张景秋

2008北京奥运 ——北京奥林匹克公园森林公园及中心区景观规划设计方案征集
International Competition for Landscaping of Forest Park and Central Zone in Olympic Green

北京市规划委员会
奥运森林公园建设管理委员会　　编
北京市园林局
中国建筑工业出版社

中国建筑工业出版社出版、发行（北京西郊百万庄）
新 华 书 店 经 销
北京广厦京港图文有限公司制作
东莞新扬印刷有限公司印刷
　　　＊
开本：965×1270毫米 1/12 印张：20⅓ 字数：976千字
2004年2月第一版　2004年2月第一次印刷
印数：1—2,800 册 定价：245.00元
ISBN 7-112-06298-5
TU·5555 (12312)

版权所有　翻印必究
如有印装质量问题，可寄本社退换
(邮政编码　100037)
本社网址：http://www.china-abp.com.cn
网上书店：http://www.china-building.com.cn

前言

北京奥林匹克公园位于北京城市市区北部，城市中轴线的北端，是举办2008年奥运会的核心区域，集中了奥运项目的大部分主要比赛场馆及奥运村、国际广播电视中心等重要设施。奥林匹克公园北部为森林公园，南部为建设区。自2001年北京成功申办奥运会以来，经过国际竞赛及后续的规划设计工作，已陆续确定了奥林匹克公园总体规划及国家主体育场、国家游泳中心等项目的设计方案。

北京城市传统的中轴线贯穿整个奥林匹克公园，在其中形成景观丰富的新城市中轴线；在森林公园中，中轴线与山形水系相结合，融入自然。

森林公园将规划建设成为一个以自然山水、植被为主的，可持续发展的生态地带，将作为北京市中心地区与外围边缘集团之间的绿色屏障部分，利于进一步改善城市的环境和气候。在奥运会举办期间，森林公园将成为奥运比赛活动的休息区。

中心区位于城市中轴线上，规划建设成为满足体育、文化、会议、商业等多功能需求的、充满活力的公共活动空间。

北京奥林匹克公园的建成，将完善北部城市功能，提升城市品质，形成新的城市形象，并加快北京向国际化大都市迈进的步伐。

受北京市人民政府的委托，北京市规划委员会、北京市朝阳区人民政府和北京市园林局于2003年7月共同组织了北京奥林匹克公园森林公园及中心区景观规划设计方案征集活动。此次征集活动面向全球范围，公开邀请优秀的、具有丰富的城市规划、景观园林规划和环保工程设计经验的设计公司、设计师事务所或设计联合体参加资格审核。申请参加资格审核的设计申请人或联合体申请人共计51个（包括了91个独立的法人实体），共涉及中国、新加坡、澳大利亚、美国、法国、德国、英国、加拿大、日本、韩国、荷兰、瑞士等12个国家和地区。在公证机构的监督下，通过专家投票选举选出了8个独立的应征人/联合体应征人。

根据征集文件的规定，主办单位邀请了30余名各专业的专家组成方案技术初审工作小组，从总体构想、景观规划、交通规划、市政规划、土方工程、造价运营等方面对所有方案进行了专业技术初审，对征集方案进行了客观的技术统计和分析，为专家评审委员会提供了技术依据。

专家评审工作历时2天，评审委员会由来自4个国家的13名委员组成，其中，国外5名，国内8名，包括规划师、风景园林师、建筑师、经济师、其他设计专家、北京市奥林匹克运动会组织委员会代表、北京市政府代表。评委最终以无记名投票的方式选出A01（易道公司、中国建筑设计研究院）、A02（Sasaki Associates, Inc.、北京清华城市规划设计研究院）、A04（北京土人景观规划设计研究所）3个方案为优秀方案。在方案评审活动结束后，主办单位组织了公开展览，同时也得到了各社会团体、广大市民和国内外友人的强烈反映，收到了许多有益的建议和意见。公众投票结果是A01、A02、A06（北京风景园林协会设计联合体）为优秀方案。

为了保留这一历史记录，对奥林匹克公园森林公园及中心区景观规划设计征集工作有一个全面、详细的介绍。本书节选了一些主要征集文件内容及所有征集方案的优秀设计，呈现给读者。

由于编者水平有限，可能其中有取舍不当之处，敬请见谅，并欢迎提出宝贵意见。

编 者
2003年12月29日

Preface

The Olympic Green located in the north part of Beijing, at the north end of the city's central axis, is the core area for hosting the 2008 Olympic Games. It includes most of the venues for the Olympic Games, and also the Olympic Village and the International Broadcast/TV Center. The north part is the Forest Park, while the south part holds constructions. Since the success of bidding for the Olympic Games in 2001, international competitions and follow-up engineering efforts have resulted in the final master planning of the Olympic Green and the architectural design of such projects as the National Stadium and the National Swimming Center.

The traditional central axis passes through the Olympic Green, where a new axis with colorful landscapes is established and integrated with hills and waters.

The Forest Park is to be developed into a natural landscape and vegetation-based ecological zone for sustainable development, as part of the green defense between the downtown area and the city fringe, to improve the local environment and climate. During the Olympic Games, it will act as the background of sports events.

The Central Zone, which is located along the central axis, is intended to be developed into a multifunctional and robust public space for sports, cultural, convention and commercial activities.

The establishment of the Olympic Green will perfect the function of the north part of Beijing, improve the city quality, create a new city image, and accelerate the pace of turning Beijing into an international metropolis.

To ensure a green landscape with biological, ecological, environmental, architectural, engineering, social and artistic elements during the 2008 Olympic Games, landscape planning of the Forest Park and the arena area in the Olympic Green was implemented as the first priority of our work in 2003. The focus of our soliciting landscape schemes was placed on how to reflect the concept of "Green Olympics, High-tech Olympics, and People's Olympics", create natural hills along the city's traditional central axis, improve the local environmental quality, and provide urban public spaces meeting the citizens' demands for recreation and leisure.

As entrusted by Beijing Municipal Government, soliciting landscape design schemes of the Forest Park and the Central Zone in the Olympic Green was jointly organized by Beijing Municipal Planning Commission, Chaoyang District Government, and Beijing Landscape Bureau, in July 2003, to invite worldwide excellent design companies, firms or joint ventures fully experienced in urban planning, landscape planning, and environmental protection engineering to prequalify for submitting landscape design schemes. Altogether 51 Participants (including 91 independent legal entities) applied for prequalification, involving 12 countries and regions, including China, Singapore, Australia, America, France, Germany, England, Canada, Japan, Korea, Netherlands, and Switzerland. As supervised by the notary public office, 8 Participants were selected by the Jury on the basis of voting.

A Technical Panel of more than 30 experts was established by the Organizer, as required in the Soliciting Documents, to perform preliminary review of the design schemes in respect of the general concept, landscape planning, traffic planning, municipal planning, soil works, construction cost, and operation. The Panel's objective statistics and analysis provided a technical basis for the Jury.

The Jury's assessment lasted for a period of 2 days. The Jury consisted of 13 members from 4 countries, 8 from China and 5 from other countries, including planners, horticulturists, architects, economists, designers, BOCOG representatives, and Beijing Municipal Government representatives. On the basis of assessment, the Jury selected 3 excellent schemes: A01 (EDAW Inc. and China Architecture Design & Research Group), A02 (Sasaki Associates, Inc. and Beijing Tsinghua Planning Corporation)、A04 (Turen Design Institute). Upon completion of the Jury's assessment, the Organizer exhibited the schemes, which have received strong response from all walks of life both at home and abroad. Also, the Organizer has received valuable recommendations and advice. Public voting resulted in 3 schemes: A01, A02 and A06 (Beijing Park Society, Beijing China Research Center of Landscape Architectural Design and Planning, Beijing Topsense Landscape & Design Co., Ltd., Tianxia Original Color Design Ltd., The Landscape Architecture School of Beijing Forestry University, Mark VanZeumeren, P.Eng.).

To maintain this historical record and to provide an all-round introduction of the soliciting, some main parts of the Soliciting Documents and all of the excellent design schemes are herein respectfully presented to our readers.

There is probably wrong choice due to our limited horizon. Your kind excuse and valuable advice would be greatly appreciated.

Editor
December 29, 2003

CONTENTS 目 录

前言
Preface

奥林匹克公园总体规划方案简介
Introduction of General Planning/Design Schemes of the Olympic Green ································ 6

奥林匹克公园森林公园及中心区景观规划设计条件
Conditions for Landscaping of Forest Park and Central Zone in Olympic Green ························ 22

奥林匹克公园森林公园及中心区景观规划设计方案征集项目情况介绍会质疑／澄清
Soliciting for Landscaping of Forest Park and Central Zone in Olympic Green Inquiry/Clarification about Project Presentation   28

奥林匹克公园森林公园及中心区景观规划设计方案
Design Schemes for Landscaping of Forest Park and Central Zone in Olympic Green
（按评审编号排序 Arranged by the Number of Assessment）································ 36

■ A01
易道公司
EDAW Inc.
中国建筑设计研究院
China Architecture Design & Research Group ······· 36

■ A02
Sasaki Associates, Inc.
北京清华城市规划设计研究院
Beijing Tsinghua Planning Corporation ········· 68

■ A03
北京市城市规划设计研究院
Beijing Municipal Institute of City Planning & Design
美国欧林景观建筑及城市设计股份有限公司
Olin Partnership Ltd. ················· 100

■ A04
北京土人景观规划设计研究所
Turen Design Institute ················ 126

■ A05
瑞驰·汉格及合作者公司
Rich Haag & Associates
艾斯弧（杭州）建筑规划设计咨询公司
XWHO (Hangzhou) Inc. ················ 158

■ A06
北京风景园林协会
Beijing Park Society
北京中国风景园林规划设计研究中心
Beijing China Research Center of Landscape Architectural Design and Planning
北京创新景观园林设计有限责任公司
Beijing Topsense Landscape & Design Co., Ltd.
北京天下原色艺术设计有限责任公司
Tianxia Original Color Design Ltd.
北京林业大学园林学院
The Landscape Architecture School of Beijing Forestry University
加拿大马克·凡泽梅伦公司
Mark VanZeumeren, P.Eng. ·············· 188

■ A07
北京市园林古建设计研究院
Beijing Institute of Landscape and Traditional Architectural Design and Research
URS 澳大利亚有限责任公司
URS Australia Pty Ltd. ················ 218

■ 后记
Postscript ······················ 244

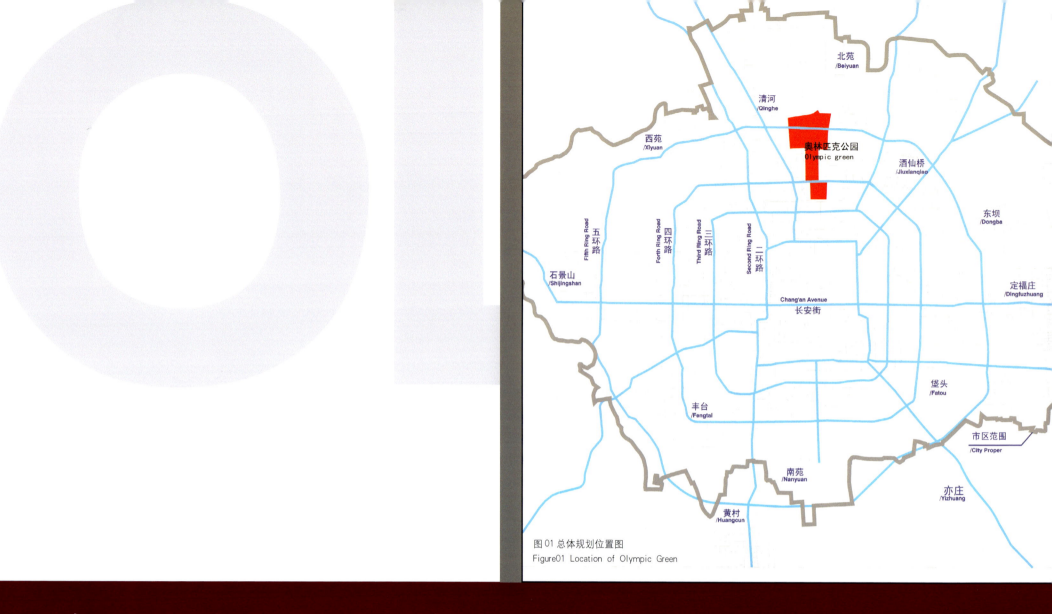

图01 总体规划位置图
Figure01 Location of Olympic Green

奥林匹克公园总体规划方案简介

奥林匹克公园总体规划方案介绍
■ 1.奥林匹克公园总体规划方案介绍
1.1 功能定位

北京奥林匹克公园位于城市中轴线的北端,是一个充满活力的、市民喜爱的、集体育、文化、展览、休闲、旅游观光为一体的多功能公共活动区域。奥林匹克公园的建成将完善北部城市功能、提升城市品质、形成新的城市形象并加快北京向国际化大都市迈进的步伐。

奥林匹克公园是举办 2008 年奥运会的核心区域,集中了 11 项比赛和 10 个场馆,并包括奥运村、国际广播电视中心、主新闻中心等重要设施。奥林匹克公园具有十分鲜明的纪念性,并成为奥运会遗产的标志。

奥林匹克公园的建设将充分利用环保的科技新成果,树立人性化设计的、可持续发展的典范。

1.2 规划指导思想
1.2.1 奥运会的建设和城市长远发展相结合。
1.2.2 延续北京城市轴线,体现文化内涵和城市建设成就。
1.2.3 体现"绿色奥运、科技奥运、人文奥运"的宗旨。
1.2.4 考虑远、近期建设结合,预留发展空间。
1.2.5 考虑融资和开发运营的可实施性。

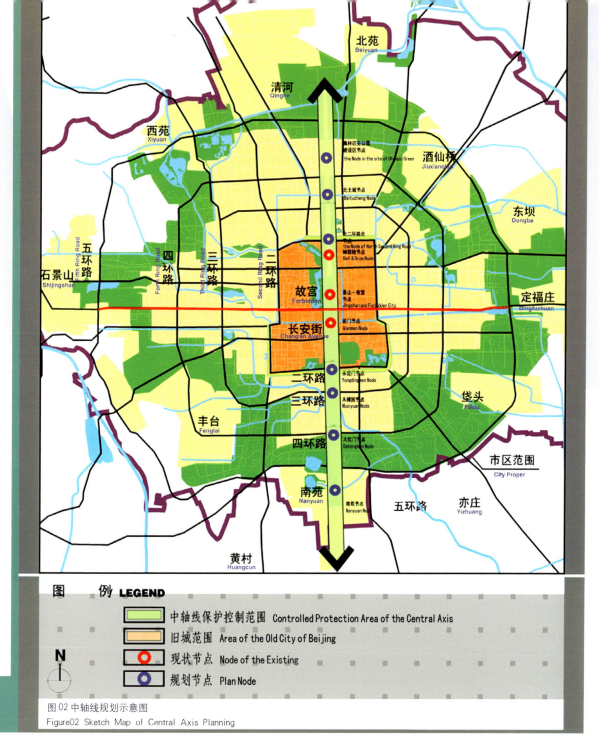

Figure 02 Sketch Map of Central Axis Planning

INTRODUCTION OF GENERAL PLANNING/DESIGN SCHEMES OF THE OLYMPIC GREEN

INTRODUCTION OF GENERAL PLANNING/DESIGN SCHEMES

■ 1. Introduction of General Planning Schemes

1.1 Functional Demands

Beijing Olympic Green (referred to as "Olympic Green" hereinafter), located at the north end of the city's central axis, is a robust and popular multifunctional public place for sports, cultural, exhibition, recreational, and sightseeing activities. Its establishment will improve the functions of the north part of the city, increase the qualities of the city, create a new city image, and speed up the pace toward an international metropolis.

The Olympic Green, as the core area for hosting the 2008 Olympic Games, will serve 11 competition events, provide 10 fields of play, and include such important facilities as the Olympic Village, International Broadcast/TV Center, and Main Press Center. It will represent an outstanding sense of souvenir and a legacy of the Olympic Games.

The Olympic Green will be constructed by making full use of up-to-date scientific and technological achievements in the field of environmental protection, to set an example of humanized design and sustainable development.

1.2 Strategies

1.2.1 Combine Olympic-oriented construction with long-term urbanization.

1.2.2 Extend the city central axis, and reflect cultural implications and urban construction achievements.

1.2.3 Reflect the aim of "Green Olympics, High-Tech Olympics, and People's Olympics".

1.2.4 Integrate long-run development with short-run construction, and with space reserved for future development.

1.2.5 Consider the achievability of financing, development, and operation.

OLYMPIC

1.3 总体规划布局（见图03奥林匹克公园总体规划及设计范围图）

1.3.1 奥林匹克公园概况

奥林匹克公园总用地约1135hm²，其中森林公园680hm²、中心区（北四环路以北）291hm²、现状国家奥林匹克体育中心用地及南部预留地114hm²、中华民族园及部分北中轴用地50hm²。

1.3.2 森林公园

森林公园位于奥运中心区北部，目前已形成较大面积的绿化。规划将尽量保留现有绿化和水面，利用工程建设的土方，堆山理水，创造景观。公园内将以自然山形水系为骨架，以常绿、落叶乔木组成的林地为基调，植被丰富，环境优美，生态良好，构成奥林匹克公园的绿色背景，充分体现"绿色奥运"的宗旨。

以水体为主要元素的龙形景观，丰富了公园景色，调节周边气候。通过湿地的设置，保证水面具备一定程度的自洁能力，为多种生物提供适宜的生存条件。水系逶迤向南，将森林公园美景引入奥运中心区。

1.3.3 中心区主要建筑布局如下（见图04中心区主要建筑功能布局图）

中心区主要包括体育设施、文化设施、会议设施和商业服务设施。

1.3.3.1 体育设施包括国家体育场（26万m²）、国家游泳中心（7.95万m²）、国家体育馆（约10.3万m²），体育设施靠近北四环路、分居于中轴线两侧，形成向南开敞的空间和较有气势的城市形象，并与南部现有奥体中心呼应。

1.3.3.2 文化设施总建筑面积（约37.64万m²），位于北辰东路的西侧，毗邻环境优美的水面和绿化，将形成北京重要的文化区域。

1.3.3.3 会议设施的定位是国家级大型国际会议中心（约33万m²），包括多个会议厅、展览厅以及附属办公、酒店等设施，位于中轴线西侧、国家体育馆北侧。展览厅在奥运会期间将作为比赛场馆，会议厅赛时用作国际广播电视中心（IBC）和主新闻中心（MPC）。

1.3.3.4 商业服务设施（约103.07万m²），位于北辰东路的西侧和北辰西路的东侧，包括商业、酒店、写字楼、娱乐、服务性公寓等内容。

1.3.3.5 奥运村（约40万m²）位于北辰西路西侧，北部为森林公园，赛后将成为一个有可持续发展示范作用的居住社区。赛时奥运村的居住区及国际区所需的临时性设施布置在其北侧的森林公园内。

1.3.4 中心区的建筑高度控制（见图05中心区主要建筑高度控制图）

除国家体育场外，奥林匹克公园建筑高度基本控制在60m以下，中轴线两侧高度为30m左右，延续北京平缓开阔的空间形态，创造优美的城市环境。

1.3.5 中心区地下空间规划（见图06、07中心区公共部分地下空间规划图）

地下空间的综合开发利用将统一考虑，将停车场及部分商业服务设施放

图03 奥林匹克公园总体规划及设计范围图
Figure03 General Layout Olympic Green and Scope of Design

1.3 General Layout (refer to Figure 03)

1.3.1 Olympic Green

The Olympic Green has a total plot area of about 1 135hm², including 680hm² for the Forest Park, 291hm² for the Central Zone, 114hm² for the existing Olympic Sports Center and the land reserved in the south, and 50hm² for the Chinese Ethnics Cultural Park and Beizhongzhou Road.

1.3.2 Forest Park

The Forest Park is located in the north part of the Central Zone, with a surface area of 680hm², including a major part of established green. It is planned to maintain the existing green and water surface wherever possible, and also create artificial landscapes by making use of soil materials resulting from project construction. Based on natural hills and water systems, a green background of colorful vegetation and beautiful environment will be created by taking forestlands of evergreen and deciduous arbors as the keynote, to fully reflect the aim of "Green Olympics".

奥林匹克公园总体规划方案简介

The dragon-shape landscape in which water is used as the main element will enrich the scenery of the Forest Park and regulate the microclimate. Wetland identified here will ensure water systems of considerable self-purifying capability, thus providing habitats for many species. Water systems winding toward the south will introduce the beauty of the Forest Park into the Central Zone.

1.3.3 Main buildings in Central Zone (refer to Figure 04)

The Central Zone will mainly include sports, cultural, convention, and commercial service facilities.

1.3.3.1 Sports facilities include the National Stadium (260 000 m²), National Swimming Center (79 500 m²), and National Gymnasium (about 103 000 m²), which are close to the North Fourth Ring Road and on both sides of the central axis. These facilities, while forming an open space toward the south and a majestic urban image, also echo with the existing Olympic Sports Center in the south.

1.3.3.2 Cultural facilities have a total floor area of 376 400 m². These facilities, on the west side of Beichendong Road immediately adjacent waters and green areas in beautiful surroundings, will be an important cultural region of the city.

1.3.3.3 The convention complex is oriented to a nationwide large-scale international convention center (about 330 000 m²), including convention halls, exhibition halls, and associated office and hotel facilities on the west side of the central axis and north side of the National Stadium. Exhibition halls will be operated as competition venues during the Olympic Games, while convention halls are used as International Broadcast Center and Main Press Center.

1.3.3.4 Commercial facilities (about 1 030 700 m²) are on the west side of Beichendong Road and on the east side of Beichenxi Road, including shopping facilities, hotels, office buildings, recreation facilities, and commercial apartments.

1.3.3.5 Olympic Village (about 400 000 m²) is on the west side of Beichenxi Road, adjacent to the Forest Park in the north. It will turn out to be a residential area as an example of sustainable development after the Olympic Games. Residential areas in the Olympic Village and temporary facilities necessary for the international zone during the Olympic Games are all provided in the Forest Park on the north side.

1.3.4 Limitation of building height in Central Zone (refer to Figure 05)

Except for the National Stadium, buildings in the Olympic Green will be essentially controlled to a height below 60 m, while those on both sides

图04 中心区主要建筑功能布局图
Figure04 Function Diagram of Main Buildings in Central Zone

图05 中心区主要建筑高度控制图
Figure05 Diagram of Main Building Height Limitations in Central Zone

入地下,以节约有限的地面空间。其中地下停车场采用集中与分散相结合的原则,车库间既相对独立设置,又设有专用公共车道将其相连,使停车设施资源共享。

将地铁站、地下商业设施连通,提高地下空间使用率,增加地下空间赢利的可能性。

奥林匹克公园中心区公共地下空间由四部分组成:

● 地铁商业部分:中轴线广场下中一路与大屯路之间结合地下一层的地铁车站布置,合理利用开挖后的地铁上部空间用作商业空间,此空间以步行商业街开发为主。

● 东侧部分:奥林匹克公园东侧,在商业开发用地与文化设施用地之间步行广场下,规划建设二层地下空间南至中一路,北至北二路,并包括与地铁车站的联系空间,此空间以步行商业街开发为主。以上两部分为中心区地下主要商业空间,总建筑面积约37.73万m²。

● 西侧隧道部分:奥林匹克公园西侧景观西路下在地下−13,000m(相对标高),设置联系隧道将景观西路两侧建设用地的地下车库联接成整体,通过信息指示系统,达到资源共享的目的,总建筑面积约3.50万m²。

● 国家体育场公共停车库部分:国家体育场西侧,景观路以东,成府路以南,在底下−7.800m(相对标高)设置地下停车库,作为国家体育场的停车场所。约1500辆停车位,总建筑面积约7.08万m²。

1.3.6 中心区地下空间及出入口控制

中轴线广场地下出入口:

中轴线广场地下出入口为解决地上与地下的方便联系及地下空间的安全疏散。地下出入口主要位于国家体育场和国家游泳馆之间公共地下车库范围内的三个下沉式广场;会议中心东侧地铁奥林匹克公园站及与其相连的部分商业空间人员进出的三个下沉式广场;发展用地东侧商业空间人员进出的三个下沉式广场。

1.3.7 中心区场地竖向规划条件

中心区竖向高程设计原则:

符合本地区基本地形走势;依据雨水、污水重力流管线控制;依据周边环境控制;中心区的竖向高程成西高东低,南高北低,且坡度平缓 (0.1%~0.2%)。

1.4 奥林匹克公园景观规划的总体设想

1.4.1 借鉴旧北京城景观设计理念,与城市整体景观融合

北京作为世界著名的历史文化名城,其旧城(明清北京城)的规划建设堪称世界都市设计的杰作。旧城以中轴线为中心、东西两侧均衡布局,在中轴线严整序列的两侧穿插了自由活泼的水系,特别是轴线西侧的三海和后海使旧城规整的布局显出生机。城市水系的相互连通和流动,实现了功能与景观的完美结合。

奥林匹克公园的景观设计借鉴了旧城传统的规划手法,由一气呵成的水系贯穿整个公园,在中轴线的东侧和北部形成湖泊,与旧城水系相呼应,使整个城市中轴线成为一个有机的整体。

1.4.2 中轴线独特的空间形态和景观

中轴线是北京旧城最重要的特色,初步形成式于元大都时期,经明清两代的发展保留至今。旧城中轴线南起永定门,北至钟鼓楼,长约7.8km,以紫禁城为中心,以景山为制高点,两侧文物建筑众多,空间收放自如,景观变化万千。

奥林匹克公园的规划注重中轴线的历史和文化的延续性,强调了公众性和景观主导性,并预留再创造余地。在北部森林公园内造湖堆山,成为轴线的背景,使中轴线融入自然山水中。

中轴线在中心区内空间形态独特,在建筑完整明确的边界之间,形成三种完全不同的空间和景观:西侧是厚实整齐的林阴道,东侧是自由舒展的水系绿化,中间是独特的中轴线空间序列。

1.4.3 创造公园独特的自然、人文环境

充分考虑北京的水源条件、气候特点以及四季景观的需求,充分利用目前已经形成的绿化及水面,引入水系,形成贯穿整个公园的景观湖面,在扩大水面的同时考虑了营造人工湿地,使未来奥林匹克公园成为生态、环保设计的典范。

保留、保护公园内的历史遗存,包括庙、墓、碑、华表等,将其与公园的景观设计整体考虑,充分发挥文物的历史价值和景观价值,并使其与公园内的景观环境融为一体。

森林公园构成了整个奥运中心区的绿色背景,为市民提供融入自然的休憩场所,增强市民的参与性。

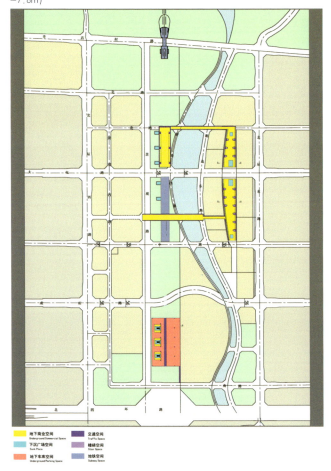

图06 中心区公共部分地下空间规划图(标高−7.8m)
Figure06 Planning Chart of Underground Public Spaces in Central Zone (Elevation −7.8m)

of the central axis are controlled to a height of about 30 m, to continue the gentle and open spatial pattern of Beijing City.

1.3.5 Underground spaces in Central Zone (refer to Figures 06 and 07)

Overall consideration will be given to the comprehensive development and utilization of underground spaces, with parking areas and some business service facilities arranged underground to save the limited ground space. Underground parking areas will be arranged on the principle of combining centralization with decentralization, with garages arranged independently and also linked by special public driving lanes, so as to share parking facilities.

Subway stations will be connected with underground business service facilities, to have a higher underground space utilization factor and a higher possibility of making profits.

The underground public space in the Central Zone is composed of 4 components:

* Subway business component: This component under the central axis square between Zhongyi Road and Datun Road will be arranged in combination with subway stations on basement 1, to make full use of the upper space of excavated subway, mainly for the development of malls.

* East component: Under the pedestrian plazas between the land for commercial development and that for cultural facilities on the east side of the Olympic Green, two-storied underground spaces will be developed up to Zhongyi Road in the south and Bei'er Road in the north, including the access space for purposes as described in Sub-section 1.2.1 above. Such spaces will be mainly used for pedestrian malls. These are the main commercial spaces in the Central Zone, with a total floor area of about 377 300 m^2.

* West tunnel component: On the west side of the Olympic Green, an access tunnel will be provided -13 m (relative elevation) underground of Jingguanxi Road, for the underground garages within the plot area on both sides of the road to be linked with each other, so as to share the resources by means of information indication systems. The total floor area here is about 35 000 m^2.

* National Stadium parking component: On the west side of the National Stadium, east side of Jingguan Road, and south of Chengfu Road, garages for 1 500 vehicles will be provided -7.8 m (relative elevation) to serve the National Stadium.

1.3.6 Control of underground spaces and exits/entries in Central Zone

Underground exits/entries at central axis square:

Underground exits/entries are provided here for easy communication between ground and underground spaces and for safety evacuation from underground spaces. They mainly include 3 sunk plazas in the range of underground public parking areas between the National Stadium and National Gymnasium, 3 sunk plazas at the Olympic Green Station and associated commercial spaces on the east side of the Convention Center, and 3 sunk plazas in the commercial spaces on the east side of land areas reserved for future development.

1.3.7 Site engineering conditions in Central Zone

Principle of vertical elevation planning:

To accord with basic topographic trends, control in accordance with gravity-current rain and sewage pipelines, control in light of surrounding circumstances, and make vertical elevations decline from west to east and from south to north at gentle slopes (0.1%~0.2%).

1.4 General Concept of Landscape Planning

1.4.1 Reference to the landscape design concept of old Beijing and integrity with the overall landscape

Beijing is a world famous historic and cultural city. Planning and construction of the old city (Beijing City of the Ming and Qing Dynasties) may be rated as a masterpiece of urban design in the world. The old city is equally distributed on both east and west sides of the central axis spiced with free and vibrant water systems on both sides, especially Sanhai and Houhai on the west side make the neat layout of the old city vivid and vital. Interconnecting and flowing water systems achieve a perfect integrity of function with landscape.

Landscape design of the Olympic Green is developed by referencing to the traditional planning techniques of the old city, with a coherent whole of water systems running through the entirety of the Olympic Green, to form lakes on the east and north sides of the central axis in response to the water systems in the old city, and make the whole central axis an organic integrity.

1.4.2 Unique spatial pattern and landscape along central axis

The central axis is the most important characteristic of the old city. Initially formed in the Yuan Dynasty, the central axis developed in the Ming and Qing Dynasties and is still preserved today. With the Forbidden City as its center and Prospect Hill as its commanding point, the central axes extends from Yongding Gate in the south and Bell and Drum Tower in the north, covering a length of about 7.8 km. A lot of buildings with cultural features stand on both sides, offering a variety of spacious and colorful views.

The Olympic Green planning lays stress on the historical and cultural continuity of the central axis and highlights the popularity and landscape predominance, with some margin reserved for further creation. Artificial lakes and hills will be created in the Forest Park in the north, to serve as the background of the central axis and integrate the axis into the natural landscape.

The central axis is of a unique spatial pattern in the Central Zone, with the following 3 totally different spaces and landscapes formed between the complete and definite boundaries of buildings: solid and neat boulevards on the west side, freely-extending water systems and green belts on the east side, and unique space series in the midst.

1.4.3 Unique physical and humane environment

2. 周边地区情况

2.1 周边地区情况

1990年第11届亚洲运动会的召开，带动了周边地区的建设，主要的公共设施现状有国家奥林匹克体育中心、国际会议中心、酒店、博物馆等，另外相关的交通设施和市政设施也比较完善。亚运会后，经过近十几年的发展，该地区已经成为北京最具吸引力的区域之一。

奥林匹克公园周围土地开发强度较高，东部有亚运村、慧忠里、安慧里和安慧北里、小关北里等居住区，西部有华严里和安翔里等居住区，总建筑量超过1000万 m²。

城市土地使用情况见下表：

用地性质	用地面积(hm²)	所占比例(%)	建筑面积(万 m²)	所占比例(%)
行政办公用地	170.52	11.09	131.58	5.98
商业金融用地	199.74	13.00	327.74	14.91
文化娱乐用地	26.72	1.74	31.37	1.43
医疗卫生用地	13.60	0.89	23.60	1.07
教育科研设计用地	156.26	10.17	200.80	9.13
宗教社会福利用地	1.45	0.09	1.95	0.09
水域	33.92	2.21	57.47	2.61
公共绿地	91.15	5.93	138.82	6.31
生产防护绿地	101.16	6.59	132.12	6.01
城市工业用地	14.70	0.96	17.62	0.80
乡镇企业	3.92	0.26	4.73	0.22
一类居住用地	0.49	0.03	0.63	0.03
二类居住用地	425.64	27.71	674.00	30.65
中学、小学、托幼用地	171.09	11.14	272.52	12.39
停车场库用地	9.31	0.61	5.21	0.24
市政公用设施用地	108.80	7.08	157.35	7.16
仓储用地	7.94	0.52	21.16	0.96
总计	1 536.41	100	2 198.77	100

上述用地均位于规划城市建设区内，另外，用地东北方向邻近的绿化隔离地区绿化用地约715hm²。

2.2 地质条件

从宏观地质条件分析，奥林匹克公园内的城市建设用地位于地形平坦的河洪冲积沉积区，地质构造条件中等，地质环境条件属简单类型，适于奥运场馆及其他各项设施的建设。

2.3 交通条件（见图08奥林匹克公园道路路网规划图）

北京城市中心区骨干路网主要由环路加放射线构成，二环路以内为旧城，四环路以内为主要的城市建设区，四环路与五环路之间规划有绿化隔离地区，奥林匹克公园即位于北四环路与北五环路之间。

目前，奥林匹克公园周边尚无轨道交通线路可达。根据城市总体规划，未来将有五条规划轨道交通线路在奥林匹克公园周边通过，分别为4号线、5号线、8号线、10号线和13号线（北京城市铁路），其中13号线2002年底已竣工，5号线正在进行试验段建设。规划设想主要利用8号线和10号线部分线位修建奥运支线到达奥林匹克公园中心区，奥运支线沿地铁8号线从奥林匹克公园向南修建到10号线，再沿10号线向东与5号线和13号线构成2个交叉换乘站。

奥林匹克公园周边有多条高速公路和城市快速路、主次干道等，与其他地区交通联系方便。

从奥林匹克公园所处位置和交通流向来看，主导交通流向为市中心方向，主要的集散道路为八达岭高速路、四环路、五环路、中轴路、安立路等，大屯路、成府路和北土城路为贯穿城市北部地区的东西向城市主次干道，也将承担部分集散交通量。

2.4 市政条件

奥林匹克公园四环路以南部分现状市政条件较好，土城北路、北四环、安立路、中轴路等道路上敷设有各种市政管线。规划新建项目的外部市政条件基本可由上述现状市政管线提供。

奥林匹克公园四环路以北部分，供水、雨水、污水、燃气等具备一定基础条件，有若干现状管线。但需要根据规划新增的建设项目，在供水、供电、供热、供气、排水、信息网络等方面还需规划建设大量新的设施和管线。如新建220kV变电站一座和110kV变电站两座；敷设综合信息管道；新建独立电信局等。

图07 中心区公共部分地下空间规划图（标高−13.000m）
Figure07 Planning Chart of Underground Public Spaces in Central Zone(Elevation−13.000m)

地下商业空间 Underground Commercial Space
楼梯空间 Stair Space
交通空间 Traffic Space
地铁空间 Subway Space

Adequate consideration is given to the local water sources, climatic characteristics, and landscape demands in all seasons in Beijing, and full use is made of established green belts and water surfaces, with water systems introduced to form waterscape through the whole Olympic Green. The development of artificial wetlands is taken into account while water surfaces are expanded, with high-tech approaches used to set the Olympic Green an example of ecological and environmental design.

Historical heritages such as temples, tombs, tablets, and ornamental columns in the Olympic Green are preserved and protected, and considered together with the landscape design as a whole, to make full use of the historical and landscape value of cultural relics and achieve an integrity with the landscape in the Olympic Green.

The Forest Park constitutes a green background of the whole Central Zone, to provide a resort for the citizens to integrate themselves into nature, and to provide them with more opportunities of participation.

■ 2. Surroundings

2.1 General

The 11th Asian Games in 1990 aroused development in the surroundings. Major public facilities include the existing Olympic Sports Center, International Convention Center, hotels, and museums. In addition, there exist fairly perfect traffic facilities and public utilities. Since then, this region has become one of the most attractive parts of Beijing over the past decade or so.

There is a high rate of land development around the Olympic Green, including the Asian Sports Village, Huizhongli, Anhuili, Anhuibeili and Xiaoguanbeili residential sites in the east, and Huayanli and Anxiangli residential sites in the west, with a total floor area over 10 million m^2.

These land uses are all within the range of planned urban construction. In addition, the adjacent green isolation region in the northeast includes some 715 hm^2 green space.

2.2 Geology

From a macro-geological point of view, the land for urban construction here is located in a fluvial sedimentary region with a flat topography, moderate geological structure, and simple geological environment suitable for the construction of facilities for the Olympic Games and other purposes.

2.3 Traffic (refer to Figure 08 for details)

The backbone road system in the city proper is mainly composed of ring roads and radial ones. The old city is within the Second Ring Road, and the main urbanization area is within the Fourth Ring Road, with green isolation areas between the Fourth Ring Road and the Fifth Ring Road, where the Olympic Green is located.

At present, there is not yet any track traffic line available to the vicinity of the Olympic Green. According to the general planning of Beijing City, however, there will be 5 track traffic lines through the perimeter of the Olympic Green, namely, line Nos. 4, 5, 8, 10, and 13 (urban railways). Among these, line No. 13 was completed in late 2002, and a test section of line No. 5 is under construction. As envisaged in the planning, an Olympic branch accessible to the Central Zone will be built mainly by taking advantage of some sections of line No. 8 and 10. This branch will be extended southward, along subway No. 8, from the Olympic Green to line No. 10, and then eastward to line No. 5 and 13, to form 2 interchange stations (see Figure).

In view of the geographic location and the traffic flow direction of the Olympic Green, the predominant flow direction is the downtown area. Major distributing roads include Badaling Expressway, Fourth Ring Road, Fifth Ring Road, Zhongzhou Road, and Anli Road. Datun Road, Chengfu Road, and Beitucheng Road, as urban main and secondary trunk roads passing through the north part of the city from east to west, will also share the traffic flow.

Land use is tabulated below:

Use	Land Arera(hm^2)	%	Floor Area(m^2)	%
Administration & office	170.52	11.09	131.58	5.98
Business & banking	199.74	13.00	327.74	14.91
Culture & recreation	26.72	1.74	31.37	1.43
Medical service	13.60	0.89	23.60	1.07
Education, scientific research & design	156.26	10.17	200.80	9.13
Religion&public services	1.45	0.09	1.95	0.09
Waters	33.92	2.21	57.47	2.61
Public green space	91.15	5.93	138.82	6.31
Protective green space	101.16	6.59	132.12	6.01
Urban industry	14.70	0.96	17.62	0.80
Township enterprise	3.92	0.26	4.73	0.22
Category A residential area	0.49	0.03	0.63	0.03
Category B residential area	425.64	27.71	674.00	30.65
School & kindergarten	171.09	11.14	272.52	12.39
Parking	9.31	0.61	5.21	0.24
Public utility	108.80	7.08	157.35	7.16
Srorehouse	7.94	0.52	21.16	0.96
Total	1 536.41	100	2 198.77	100

图08 奥林匹克公园道路路网规划图
Figure08 Planning Chart of Road Networks in Olympic Green

■ 3. 规划设计建筑方案介绍

3.1 国家体育场简介

3.1.1 概况

国家体育场是奥林匹克公园的标志性建筑物之一，是2008年北京奥运会的主体育场，将承担开闭幕式和田径比赛，赛后可用于各种国际水平的田径、足球及其他表演活动。它位于中心区中轴线东侧、龙形景观水系的西侧。同国家游泳中心、国家体育馆分列于中轴线两侧。

在奥林匹克公园总体规划方案确定后，2002年底，北京市政府启动了国家体育场（2008年奥运会主体育场）建筑概念设计竞赛工作。瑞士Herzog & De Meuron建筑设计公司和中国建筑设计研究院合作的方案被选定为实施方案。国家体育场基地占地面积：约300m×260m，并附带一个400m跑道的永久性热身场地，总建筑面积约26万m²，永久坐席8万人，临时坐席2万人，屋面部分可开启，建筑高度约68m。该方案的设计深化工作正在进行。工程项目计划于2003年年底开工。

3.1.2 设计特点

体育场好像一个巨大的容器，建筑形体单纯完整，结构构件组成建筑外观。国家体育场主要结构由巨大的门式钢桁架沿椭圆辐射状旋转形成主受力体系，中间沿传力路径填充次级构造钢架，将看似无序的框架纳入严谨的受力体系中，结构功能与建筑外观统一，秩序蕴涵着变化。

屋面和部分构架单元之间以ETFE充气膜填充。进入体育场的界面为开敞式设计，不设立封闭的实墙，空间开

奥林匹克公园总体规划方案简介

INTRODUCTION OF GENERAL PLANNING/DESIGN SCHEMES OF THE OLYMPIC GREEN

2.4 Public Utilities

There are favorable public utilities south of the Fourth Ring Road in Olympic Green, with various types of municipal pipelines laid along Tuchengbei Road, North Fourth Ring Road, Anli Road, and Zhongzhou Road. Proposed external public utilities can be satisfied essentially by these existing pipelines.

There exist some water supply, rain, sewage, and gas infrastructures in the north of the Fourth Ring Road, with sever pipelines available there. However, large numbers of water supply, power supply, heat supply, gas supply, drainage, and information service facilities and pipelines will be provided according to the demands of proposed new projects, e.g., a 220kV substation and two 110kV substations, integrated information pipes, and an independent telecommunication office, etc.

■ 3.Introduction of Architectural Schemes
　3.1 National Stadium
　3.1.1 General

The National Stadium is one of the landmarks in the Olympic Green. As the main stadium for the 2008 Olympic Games, it will witness the opening and closing ceremonies, tract and field events, as well as various international track and field events, football matches, and other entertainments after the Olympic Games. It is on the east side of the central axis through the Central Zone and both sides of the dragon-shape water system. The National Stadium, National Swimming Center, and National Gymnasium are distributed on both sides of the central axis.

With the general planning of the Olympic Green determined, competition for the architectural concept design of the National Stadium was initiated by the Beijing Municipal Government in late 2002. The scheme jointly developed by Herzog & De Meuron (Swiss) and the China Architectural Design Institute is determined as the implementation scheme. The National Stadium has a plot area of about 300m x 260 m, and additionally a permanent warm-up field with 400 m lanes, with a total floor area of some 260000 m^2, a total capacity of 80000 permanent seats plus 20 000 temporary ones, a partly retractable roof, and a building height of about 68 m. In-depth design is being carried out. The project is scheduled to be started by the end of 2003.

　3.1.2 Design characteristics

The National Stadium looks like a huge container. The architectural shape is pure and complete, with architectural appearance constituted by structural components. The main structure is composed of huge gate-type steel frames rotating around an oval in the radial direction, to form the main bearing system. Sec-

奥林匹克公园总体规划方案简介

放,并确保最大限度的自然通风。

设计者将体育场外部地面缓缓隆起约4m,形成体育场地的基座,基座下方的空间可根据运营的需要灵活安排各种体育比赛需要的设施或商业运营空间。观众可直接进入体育场中部看台。室外地面刻有十二生肖图案,划分出体育场的不同入口区域。

3.1.3 功能分区

地下二层为停车库、竞赛用房、体育场附属设施。竞赛场地在此层。

地下一层安排有部分地下商业设施。主体育场通过地下通道或地面天桥,和热身场地相连。

地上一层为入口大厅,高于中心区地平4m。观众由此进入中部看台。

上部各层均位于看台下。各种用房主要满足比赛需要,并可用于赛后经营。

3.1.4 奥运会期间的使用要求

奥运会期间,需要在场馆周边部分提供临时性赛事组织和管理的临时用地,包括官员、贵宾、记者的专用车辆停车场及其他临时设施。这些设施和用地在国家体育场用地范围内解决。

3.2 国家游泳馆简介

3.2.1 概况

国家游泳中心位于北京奥林匹克公园中心区的南部,是为举办2008年奥运会新建的三大场馆之一,规划建设用地约6.295hm²,主体建筑紧邻北京城市中轴线,并与拟建的国家体育场相对于中轴线均衡布置。

国家游泳中心将设计建设成国际先进水平的大型水上运动中心。2008年奥运会期间将作为游泳、跳水、花样游泳、水球等比赛用馆;奥运会赛前及赛后将作为一个多功能的大型水上运动中心,为公众提供大型的多功能水上娱乐、运动、休闲、健身场所,还可举办国际、国内游泳、跳水体育赛事。

通过方案竞赛,中建、PTW、ARUP联合体设计的"水立方"方案被选中作为实施方案。该方案总建筑面积约80 000m²(地上约50 000m²,地下约30 000m²),奥运会期间坐席总数共计17 000个,包括永久坐席6 000个,临时坐席11 000个建筑高度:约35m,基地占地面积:约34 000m²。工程项目计划于2003年年底开工。

3.2.2 设计特点

建筑师从"水"与"方"这两个概念入手。通过模仿和抽象水泡结构形式,形成国家游泳中心独特的结构体系,并利用ETFE气囊作为建筑外立面材料,形成水泡的肌理,体现国家游泳中心作为水上项目的功能特点。设计者利用完整的"方"型同国家体育场"椭圆"外观的呼应关系,形成水与火,放射与内敛的不同性格,同时,

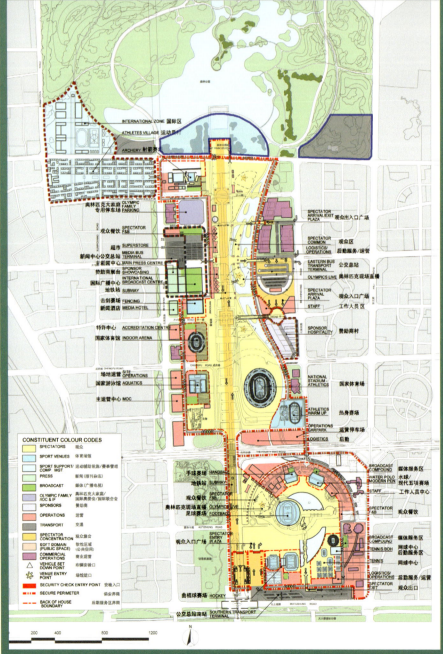

图9 奥林匹克公园赛时组织示意图
Figure09 Conceptual Operational Plan of Olympic Green during the Olympic Games

ondary structural steel frames are arranged along the path of transmission, which incorporates the frames that appear disorderly into the well-knitted bearing system, integrates structural functions and architectural appearance, and implies varying orders.

ETFE pneumatic membranes are placed between the roof and some structural units. The interface into the stadium is designed to be open, without any enclosed solid wall, to provide open space and enable maximum natural ventilation.

The ground outside the stadium is gently raised about 4 m, to form a foundation bed. In the space below the foundation bed, facilities necessary for sports events or commercial spaces can be arranged flexibly below the foundation bed. Spectators can get to the middle stands directly. Patterns of the twelve animals (used to symbolize the year in which a person is born) are provided outside, to identify the entry areas.

3.1.3 Sectorization

Basement level 2 includes garages, event management rooms, and associated facilities, and field of play.

Basement level 1 includes some commercial facilities. The stadium is accessible to the warm-up area via underground accesses or overline bridges.

Ground level 1 is the entry concourse, 4 m higher than the ground in the Central Zone. Spectators will get into the middle stands from here.

Upper levels are all below the stands. Rooms in different types are provided to mainly meet the demands of events, but they will also operated for commercial purposes after the Olympic Games.

3.1.4 Operational demands during the Olympic Games

During the Olympic Games, land areas for the temporary use of event operation and management will be required in the surroundings, including special parking areas and other temporary facilities for the use of officials, VIPs, and journalists. These will be identified in the plot area of the National Stadium.

3.2 National Swimming Center

3.2.1 General

The National Swimming Center located in the south part of the Central Zone is one of the 3 major venues to be provided for the 2008 Olympic Games. It is planned to have a plot area of about 6.295 hm^2. The main building is immediately adjacent to the central axis, arranged together with the proposed National Stadium symmetrically about the central axis.

The National Swimming Center is to be designed and built as a large aquatic sports center up to the world's best practice. It will be used for swimming, diving, synchronized swimming, and water polo events during the 2008 Olympic Games, and operated as a large multifunctional aquatic sports center before and after the Olympic Games, to provide the public with a space for aquatic recreation, sports, leisure, and body-building activities. Also, it will be capable of serving international and national swimming and diving events.

Through competition for architectural design, the "water cube" scheme developed jointly by China Architectural Design Institute, PTW, and ARUP is selected as the implementation scheme. This scheme proposes a total floor area of about 80 000 m^2 (including some 50 000 m^2 ground and 30 000 m^2 underground), total capacity of 17 000 seats during the Olympic Games, including 6 000 permanent and 11 000 temporary ones, building height of about 35 m, and plot area of about 34 000m^2. Commencement of the project is scheduled to happen in late 2003.

3.2.2 Design characteristics

Starting from the concept of "water" and "square", the designer develops a unique structural system by simulating and abstracting the structural style of bubbles. ETFE balloons are used as external elevation material, to form the organisms of bubbles, and reflect the functional characteristics of the National Swimming Center as an aquatic project. A complete "square" responding to the "oval" appearance of the National Stadium is utilized to provide the radiating and restricting characters of water and fire. Also, the architectural design contains some traditional cultural ideas of China.

The National Swimming Center provides a greatly varying background for Beiding Goddess Temple.

INTRODUCTION OF OLYMPIC DESIGN SCHEMES O

也在建筑设计中蕴涵着一些中国传统的文化理念。

国家游泳中心为北顶娘娘庙提供了一个多变的背景。

国家游泳中心室外将设计一系列圆形水池,寓意方块落在基地溅出的水花,水滴四溅,滴水荡出涟漪。这些"水滴"形成场地上与水相关的景观小品,如水池(有选择地种以植物)、雕塑、喷泉等。国家游泳中心四周以线性水池与陆地分开,正如中国古城池被河水环绕一样,以桥连接建筑入口处。

3.2.3 建设内容

主要为水上运动项目,包括:两个标准游泳池,保证所有水上运动项目的需求,赛后副场地上加建一层,改造成为多功能馆;一个跳水池,一个水上娱乐中心。

配套经营活动包括:办公、商业、体育俱乐部和文化设施。

3.2.4 奥运会期间的使用要求

奥运会期间,需要在场馆周边部分提供临时性赛事组织和管理的临时用地,包括官员、贵宾、记者的专用车辆停车场及其他临时设施。这些设施和用地在国家游泳中心用地范围内解决。

3.3 会议中心简介

3.3.1 概况

会议中心位于奥林匹克公园中心区的中部,城市中轴线西侧,规划建设用地约12.21hm^2,总建筑规模约350 000m^2,平均容积率小于3,建筑控制高度为30~45m。会议中心东侧为200m宽的中轴线步行绿化广场,西侧规划建筑控制高度为60m的商业开发用地(包括酒店、公寓、办公等),南侧规划为国家体育馆,北侧规划为发展预留用地。

会议中心是北京最重要的、具有国际先进水平的多功能会议中心,适合举办各种类型的国际会议,并可以提供多功能展示的场所。它由会议、展览、餐饮、办公、酒店、商业零售、其他辅助活动以及后勤服务设施等内容构成,能够持久地为地区经济活动提供动力。其中会议、展览与之配套的餐饮功能为会议中心的核心功能;办公、酒店和商业零售为会议中心的必要功能;其他辅助活动和后勤服务设施为会议中心的支撑功能。会议中心将成为奥林匹克公园中的重要建筑。

3.3.2 建设内容

3.3.2.1 功能分区

景观路和景观西路之间用地,建设会议及配套服务设施(会议、展览、商业零售等),建筑总规模约150 000m^2。

景观西路和北辰西路之间用地,建设会议配套设施(酒店、公寓、办公等),建筑规模约200 000m^2。

3.3.2.2 面积指标

会议及餐饮设施:建筑面积约40 000~70 000m^2。

展览设施:总建筑面积约40 000m^2。

酒店:配套建设部分3~5星级酒店,提供900间标准客房的住宿能力,每间净面积约20~40m^2。

酒店式公寓:总建筑面积约30 000m^2。

零售商业:为配合会议中心的运营需要,需设置商业零售面积约40 000m^2。主要商业面积应沿景观路一侧连续布置。

办公设施:供会议、展览的组织者或其他相关人员提供可出租的办公楼,建筑面积约20 000m^2。

其他辅助面积和后勤场地:不超过50 000m^2。

3.3.3 奥运会期间的使用要求

3.3.3.1 功能要求

奥运会期间,会议中心将提供一个临时比赛场馆,座位数为10 000座,作为击剑比赛场地、现代五项比赛中击剑和气手枪比赛场地,同时要为奥运会临时提供国际广播电视中心(IBC,约为83 500m^2)和主新闻中心(MPC,58 400m^2)的用房。

3.3.3.2 奥运会期间"后院"的要求

GENERAL PLANNING / THE OLYMPIC GREEN

A series of round ponds are provided outdoors, to imply water blooms splashed by squares falling on the foundation and subsequent water drops splashing and rippling in all directions. These "water drops" form small landscape creations in association with water, such as ponds (plants selectively), sculptures, and fountains. Linear ponds are provided around the National Swimming Center, to separate it from the land, just like Chinese ancient cities surrounded by waters, with bridges provided to building entries.

3.2.3 Components

Main aquatic facilities include 2 standard swimming pools meeting the demands of all aquatic events (rebuilt into a multifunctional natatorium by adding a level to the secondary field after the Olympic Games), a diving pool, and an aquatic recreation center.

Supporting business operations include office, commercial, sports club, and cultural facilities.

3.2.4 Operational demands during the Olympic Games

During the Olympic Games, land areas for the temporary use of event operation and management will be required in the surroundings, including special parking areas and other temporary facilities for the use of officials, VIPs, and journalists. These will be identified in the plot area of the National Swimming Center.

3.3 Convention Center

3.3.1 General

The Convention Center is located in the middle of the Central Zone, on the west side of the central axis, with a plot area of about 12.21 hm^2, total floor area of about 350 000 m^2, average plot ratio of less than 3, and building height of 30~45 m. On the east side is a 200 m wide central axis square, on the west side is land for commercial development (including hotel, apartment, and office building, etc.) where there is a planned building height limitation of 60 m, on the south side is the National Stadium, and on the north side is land reserved for future development.

The Convention Center, the most important multifunctional convention center up to the world's best practice in Beijing, is suitable for various international conventions and provides multifunctional exhibition places. It consists of convention, exhibition, catering, office, hotel, retail, and other auxiliary and logistic facilities. Offices, hotels, and retail facilities are essential functions, while other auxiliary activities and logistic facilities are supporting functions. The Convention Center will be an important building in the Olympic Green.

3.3.2 Components

3.3.2.1 Sectorization

Between Jingguan Road and Jingguanxi Road are convention and supporting facilities (convention, exhibition, and retail facilities) with a total floor area of some 150 000 m^2.

Between Jingguanxi Road and Beichenxi Road are supporting facilities (hotel, apartment, and office) with a total floor area of about 200 000 m^2.

3.3.2.2 Area indexes

Convention and catering facilities: a floor area of about 40 000~70 000 m^2;

Exhibition facilities: a total floor area of about 40 000 m^2;

Hotels: some 3~5-star hotels with a total capacity of 900 standard rooms, 20~40 m^2 each;

Service apartments: a total floor area of about 30,000 m^2;

Retail facilities: a floor area of about 40 000 m^2, with main commercial areas continuously arranged on a side of Jingguan Road;

Office facilities, a floor area of about 20 000 m^2, with office buildings provided for rent by convention and exhibition organizers or others as may be concerned;

Other auxiliary areas and logistic spaces: not more than 50 000 m^2.

3.3.3 Operational demands during the Olympic Games

3.3.3.1 Functional demands

During the Olympic Games, the Convention Center will provide a temporary venue with a capacity of 10 000 seats, to serve fencing, and fencing and air pistol as part of modern pentathlon. Also, rooms will be made available for an international broadcast center (about 83 500 m^2) and a master press center (58 400 m^2).

3.3.3.2 Demands of "back of house"

Before the Olympic Games, construction will not be carried out in some spaces, which will be used as the back of house (for outdoor operation of the international broadcast center and the master press center, and for the use of parking and security check), with construction performed as planned thereafter.

3.4 National Gymnasium

3.4.1 General

The National Gymnasium is in the south part of the Central Zone, with a plot area of about 6.7 hm^2. The main building is immediately adjacent to the central axis, arranged together with the National Stadium symmetrically about the central axis. The building height is controlled below 45 m, and percentage of green spaces is not lower than 30%. On the east side is a 200 m wide central axis square, on the west side are commercial facilities with a planned building height of 45~60 m (including hotels, and office buildings), on the south side is the National Swimming Center, and on the north side is the proposed Convention Center.

3.4.2 Components

The National Gymnasium is one of the 3 major sports buildings in the Olympic Green. With a total capacity of 19 000 seats, including 16 000

奥林匹克公园总体规划方案简介

奥运会前,部分用地不得建设,作为体育场赛时后院(IBC、MPC所需室外运行和停车安检场地),赛后方可按规划建设。

3.4 国家体育馆简介

3.4.1 概况

国家体育馆位于奥林匹克公园中心区的南部,规划建设用地约6.7hm²,主体建筑紧邻城市中轴线,并与国家体育场相对于中轴线均衡布置,建筑高度控制在45m以下,绿化率不少于30%。国家体育馆东侧为200m宽的中轴线步行广场,西侧规划建设高度为45~60m的商业设施(包括酒店、办公等),南侧规划为国家游泳中心,北侧规划为会议中心。

3.4.2 建设内容

国家体育馆是北京奥林匹克公园内的三大重要场馆之一,总座位数19 000个,其中永久性座位16 000个,临时性座位3 000个。建筑面积约67 000m²,是北京最大的、具有国际先进水平的多功能运动、健身、休闲中心。国家体育馆将成为奥林匹克运动留给城市的宝贵遗产和城市建设的新亮点。

国家体育馆将承担体操、排球、手球等奥运会比赛。奥运会后可承担特殊重大比赛(如:残奥会、世界体操锦标赛)、各类常规赛事(如:亚运会、洲际综合性比赛、全国运动会等),成为一个集比赛、训练、娱乐、健身于一体的综合室内运动中心。

3.4.3 奥运会期间使用要求

奥运会期间,奥林匹克公园赛时实行分级封闭管理,观众的安检口分别设在中心区的北端和南端。国家体育场周边还将设置"赛时后院"。(是指奥运会赛时组织和管理的临时用地,包括官员、贵宾、记者的专用车辆停车场及其他临时设施)。后院用地在国家体育馆西南侧用地内安排,赛时将用于奥运会组织、管理和媒体所需的功能。

目前,国家体育馆项目的设计方案尚未确定。

MPIC

INTRODUCTION OF GENERAL PLANNING/DESIGN SCHEMES OF THE OLYMPIC GREEN

permanent and 3 000 temporary ones, and a floor area of about 67 000 m², it is the largest internationally advanced multifunctional sports, bodybuilding, and leisure center in Beijing. It will be a valuable legacy bequeathed by the Olympic Games in Beijing and a new sense of pride in the field of urban construction.

The National Gymnasium will serve gymnastics, volleyball, and handball events during the Olympic Games, and also peak events (for example, the Paralympic Games, World Gymnastics Championships) and various conventional events (for example, the Asian Games, continental games, and national games) thereafter, to become a multifunctional indoor sports center for competition, training, recreation, and bodybuilding.

3.4.3 Operational demands during the Olympic Games

During the Olympic Games, enclosed management by levels will be applied to the Olympic Green, with spectator security checks provided at the north and south ends of the Central Zone respectively. Enclosed secondary areas will be identified around the National Gymnasium, to ensure spectators of easy access and avoid crossing of coming and leaving circulation routes. In addition, "back of house" (refer to spaces for temporary use of event operation and management, including special parking and other temporary facilities for the use of officials, VIPs, and journalists) will be provided on the perimeter of the National Stadium. Such spaces will be identified in the southwest of the plot area of the National Stadium, to meet event operation, management, and media demands during the Olympic Games.

At present, the design scheme of the National Gymnasium is not determined.

■ 1. 项目名称
奥林匹克公园森林公园及中心区景观规划设计（以下简称森林公园及中心区景观规划设计）

■ 2. 位置及范围

2.1 位置
奥林匹克公园森林公园及中心区位于北京市区北部、城市中轴线的北端（见图01 奥林匹克公园位置示意图）。

2.2 范围（见图03 奥林匹克公园总体规划及设计范围图）

2.2.1 森林公园及中心区景观规划设计的用地范围包括
奥林匹克公园森林公园：约700hm^2（含清河河北以南地区）
奥林匹克公园中心区：约160hm^2
2.2.2 扩大研究范围：约80hm^2
2.2.2.1 元大都城垣遗址公园北部部分地段（以下简称元大都城垣遗址公园）：29hm^2
2.2.2.2 现状中轴路红线两侧50m范围、北四环路红线以南50m范围用地：14.4hm^2
2.2.2.3 国家体育场用地：20.4hm^2
2.2.2.4 清河北岸70m河道绿化带用地：16.2hm^2

■ 3. 景观规划设计的目的
3.1 根据已确定的奥林匹克公园总体规划，完善奥林匹克公园景观设计。
3.2 统一考虑森林公园、中心区两个主要设计范围，并区分两个地区不同的功能要求：把握森林公园作为北京市区中心区和外围边缘集团之间绿色屏障的城市功能定位，将森林公园建成一个以自然山水、植被为主的，可持续发展的休憩空间；结合中心区规划布局、交通组织、主要建筑方案，将中心区建设成为满足体育、文化、会议、商业等多功能需求的、充满活力的公共活动空间。
3.3 统一考虑长期利用和奥运会比赛组织要求，为成功举办奥运会创造条件。
3.4 为市民创造满足多种休憩需要的城市公共活动空间，改善城市环境及小气候。

奥林匹克公园森林公园及中心区景观规划设计条件

CONDITIONS FOR LANDSCAPING OF FOREST PARK AND CENTRAL ZONE IN OLYMPIC GREEN

■ 1. Project Name

Landscaping of Olympic Forest Park and Central Zone (refer to as landscaping of Forest Park and Central Zone hereinafter)

■ 2. Location and Plot Area

2.1 Location

The Forest Park and the Central Zone area is located in the north part of Beijing, at the north end of the city's central axis (refer to Figure 01).

2.2 Plot Area (refer to Figure 03).

2.2.1 The plot area includes:

Olympic Forest Park: about 700hm^2 (including the area south of Qinghe River)

Central Zone: about 160 hm^2

2.2.2 Study Area: about 80 hm^2

2.2.2.1 Some areas in the north of the Park of the Historical Heritage of the Capital Wall of Yuan Dynasty (referred to as Park of the Historical Heritage of the Capital Wall of Yuan Dynasty): 29 hm^2

2.2.2.2 The areas 50 m within the property line on both sides of Zhongzhou Road, and the area 50m within the property line to the south of the Fourth Ring Road: 14.4 hm^2

2.2.2.3 National Stadium: 20.4 hm^2

2.2.2.4 The 70 m greenbelt on the north bank of the Qinghe River: 16.2 hm^2

■ 3. Purpose

3.1 Improve the Olympic Green landscape design on the basis of the established general planning;

3.2 Give overall consideration to the Forest Park and the Central Zone, but distinguish their separate functional demands; develop the Forest Park into a resort for sustainable development by mainly relaying upon natural landscape and vegetation, and keeping in mind its function that it will act as a green defense between the downtown area and the fringe of Beijing City; and build the Central Zone into a multifunctional and robust public space for sports, cultural, convention, and commercial activities in consideration of the general layout, traffic arrangements, and major architectural schemes;

3.3 Consider the demands of long-term operation and those of Olympic event operation as a whole, to create favorable conditions for the success of the Olympic Games;

3.4 Plan and develop an important urban space and environment in the north part of the city, to provide the citizens with a multifunctional public space, and to improve the local environment and microclimate.

奥林匹克公园森林公园及中心区景观规划设计条件

■ **4. 景观规划设计的指导思想**

4.1 充分考虑城市中轴线在奥林匹克公园部分的景观处理
北京城市传统的中轴线贯穿整个奥林匹克公园,并在城市北部形成新的城市中心区。因此景观规划设计应尊重原有城市轴线,并结合奥林匹克公园总体规划中入口广场、体育广场、交通广场、文化休闲广场的设计,形成景观丰富的新城市轴线;在森林公园,使中轴线与山形水系相结合,融入自然。

4.2 符合生态环境要求,有利于改善和提高生态环境质量
通过保护和改善原有生态环境,结合人造景观环境的设计和对地方原生物种的保护,保持原有自然生态系统的均衡,同时运用高科技的环保生态技术及材料,进一步提高城市生态环境质量,形成环境的可持续发展。

4.3 考虑公园长远规划与奥运短期服务的双重需要
森林公园及扩大研究范围将作为城市中的自然生态公园,中心区及扩大研究范围将成为多功能的市民活动中心,共同构成奥林匹克公园的内容,成为城市北部开放的大型市民公共活动空间。景观规划设计应在充分满足城市长远规划的同时,为奥运会的使用要求留有充分的余地。

4.4 体现"绿色奥运、科技奥运、人文奥运"的宗旨
规划设计方案应体现"绿色奥运、科技奥运、人文奥运"的宗旨。应有利于环境保护和可持续发展,采用当今国际上最新的规划设计理念和先进的技术手段,提高生态环境质量;广泛普及奥林匹克精神,弘扬中华民族优秀传统文化,展现北京历史文化名城风貌,建立与举办奥运会相适应的人文环境。充分考虑各类人员的需要,尤其要关注残疾人和有行动障碍人员的需求等。

CONDITIONS FOR LANDSCAPING OF FOREST PARK AND CENTRAL ZONE IN OLYMPIC GREEN

■ 4. Strategy

4.1 Give Full Consideration to Landscape Treatment of the Central Axis

The city's traditional central axis passes through the entirety of the Olympic Green and forms a new downtown area in the north of the city. In this instance, landscaping shall be developed to show respect for the existing axis and to establish a new axis of rich landscapes in consideration of the entry, sports, traffic, and recreation plazas proposed in the general planning of the Olympic Green, for the central axis to be integrated into nature in the Forest Park.

4.2 Comply with Environmental Requirements, Facilitate Environmental Improvement

Local protist species are protected and existing ecosystem balance is maintained through environmental protection and improvement in combination with artificial landscaping. In furtherance, high-tech environmental-friendly technologies and materials are applied for further improvement of environmental quality and sustainable development of environment.

4.3 Consider Long-term Planning and Short-term Service

The Forest Park will be regarded as a natural ecological park in the city, whereas the Central Zone and the Study Area (south of Zhongzhou Road-Xiongmaohuandao) are operated as a multifunctional public center, to jointly constitute the Olympic Green and provide a large-scale open public space in the north part of the city. While the long-term planning of the city is fully satisfied, an adequate margin should be considered for meeting the demands of the Olympic Games.

4.4 The landscaping scheme is required to reflect the aim of "Green Olympics, High-tech Olympics, and People's Olympics": facilitate environmental protection and sustainable development, achieve environmental improvement by applying the most up-to-date design concepts and techniques, spread the spirit of the Olympics, promote the excellence of the traditional Chinese culture, show the new look of Beijing as a famous historical and cultural city, establish a people's environment suitable to the Olympic Games, and given adequate consideration to the demands of different categories of people, especially those with physical disability.

奥林匹克公园森林公园及中心区景观规划设计条件

■ 5. 景观规划设计的结构性要素

5.1 中轴线

北京城的中轴线贯穿整个奥林匹克公园,将在中心区内达到空间序列的高潮,并将继续向北延伸至森林公园后渐入自然山水之中。因此规划要着重研究城市中轴线与奥林匹克公园的关系,处理好中轴线与各景观要素的衔接,恰当地体现中轴线在北京城市空间和自然背景中的独特效果,组织好城市景观空间的序列。

5.2 山形水系

山体主要分布在森林公园。水面与山体有机结合,构成该区域良好的自然山水环境。

奥林匹克公园规划水系贯穿整个奥林匹克公园,主水面及湿地安排在森林公园内五环路以南地区的中央。整个水系将成为奥林匹克公园内城市轴线上的活跃要素,为公园提供了优美的城市景观及环境。

因此,景观规划应处理好山形水系与中轴线、绿化、广场等要素的关系,充分发挥山水要素在城市中的亲和作用以及生态环保效果。

5.3 重要节点

根据奥林匹克公园总体规划,依据景观规划用地周围规划建筑的功能要求及性质,整个奥林匹克公园景观规划沿中轴线由南向北依次形成五个重要节点。应征人可根据自身理解和方案情况确定其他重要节点。

5.3.1 入口区:入口区位于中心区规划范围南段、中轴线上,紧临北土城路元大都城垣遗址公园,是进入奥林匹克公园的空间序曲。因此景观规划应突出进入奥林匹克公园入口的标志性形象,同时处理好与遗址公园的衔接。

5.3.2 体育区:体育区位于奥林匹克公园中心区南端,东侧为国家体育场,西侧为国家体育馆和国家游泳中心,形成体育建筑群。因此景观规划应突出体育区的特色,并且满足大型体育活动对人员疏散的要求。

5.3.3 交通广场区:交通广场区位于奥林匹克公园中心区中部,设有中心区内地铁奥运支线惟一的站点,是人员的主要聚散点。因此景观规划应结合人员疏散功能,创造丰富的硬质景观广场,形成整个奥林匹克公园中心区景观视觉中心。

5.3.4 文化休闲区:文化休闲区位于奥林匹克公园中心区北部,东侧为文化设施,西侧为国际会议中心,形成以文化设施为主的建筑群。因此景观规划应突出文化特色,结合景观规划设计,形成市民休闲活动区。

5.3.5 森林公园主入口区:森林公园主入口区位于奥林匹克公园中心区北部、森林公园南部,是两个区域景观的衔接部分,同时也是两个区域的主要出入口。因此景观规划应突出整体景观的连贯性,同时标识出两大区域各自的特色。

5.4 标志性景观

奥林匹克公园是北京城市北部地区及北中轴线上的重要城市开放空间,建议在景观规划设计范围内,规划设计能够统率整个奥林匹克公园的标志性景观。位置、构思应由设计人结合方案设计完整表达。

5. Structural Elements of Landscaping

5.1 Central Axis

The central axis passes through the whole Olympic Green, reaches the climax of the serial spaces in the Central Zone, and then extends northward to the Forest Park where it integrates itself into natural landscape. Therefore, the focus of landscaping should be placed on the following aspects: studying its relationship with the Olympic Green, properly treating its linkage with various landscape elements, appropriately reflecting its unique effect on the urban space and natural background of Beijing, and properly arranging the series of urban landscape spaces.

5.2 Water Systems

Water systems are planned to run through the entirety of the Olympic Green, with main water systems and wetlands arranged in the center of the Forest Park south of the Fifth Ring Road. These water systems will be a lively element of the central axis through the Olympic Green and provide a nice urban landscape and environment. Thus, landscaping should properly treat their relationship with the central axis, green spaces, and plazas among other elements, and make full use of waterscape affinity for the city and its effect of environmental protection.

5.3 Important Nodes

According to the general planning of the Olympic Green, and based on the functions and nature of proposed buildings in the plot area, the whole landscaping includes 5 important nodes along the central axis from south to north. The Participant may identify other important nodes according to his understanding and the schemes.

5.3.1 Entry area: This area is at the south end of the Study Area (south of Zhongzhou Road to Xiongmaohuandao), immediately adjacent to the Park of the Historical Heritage of the Capital Wall of the Yuan Dynasty. It is the space overture of the Olympic Green. Thus, the landscaping shall highlight the symbolic image of the entry and properly handle its relationship with the Park of Historical Heritage of the Capital Wall of the Yuan Dynasty.

5.3.2 Sports area: This area is at the south end of the Central Zone, with a sports building complex formed by the proposed Station Stadium in the east and the National Gymnasium and the National Swimming in the west. The landscaping shall highlight the characteristics of this sports area and meet the requirement for evacuation during large-scale sports events.

5.3.3 Traffic plaza area: This area is in the middle of the Central Zone. The only substation in the Central Zone is located in this area, so it is the major source of people. The landscaping shall create plenty of hardened landscape plazas in light of evacuation functions, to form a center of landscape version in the whole Central Zone.

5.3.4 Cultural and recreational area: This area is in the north part of the Central Zone, with a building complex formed by predominant cultural facilities in the east and the International Convention Center in the west. The landscaping shall lay stress on the cultural characteristics, to provide a recreation space for the citizens.

5.3.5 Main entry area of Forest Park: This area is between the Central Zone and the Forest Park, acting as linkage between and a main entry into these two parts of the Olympic Green. The landscaping shall highlight the continuity of the whole landscape and identify the separate characteristics of these two parts.

5.4 Symbolic Landscape

The Olympic Green is an important open urban space in the north part of Beijing and on the north central axis. It is herein recommended to provide a symbolic landscape that will govern the whole Olympic Green, with the location and concept adequately expressed by the Participant in consideration of the schematic design.

OLYMPIC

奥林匹克公园森林公园及中心区景观规划设计方案征集项目情况介绍会质疑／澄清

在本次征集活动中，主办单位向应征人提供的所有资料（包括但不限于文字资料、图纸、图片、照片、电子文件等等）的知识产权尤其是版权均归主办单位或其他的版权人所有。任何应征人不得将该资料提供给任何第三方使用，且不得用于除准备应征奥林匹克公园森林公园及中心区景观规划设计方案之外的任何目的。

问题1

能否提供奥林匹克公园的地质勘探资料，以及用地范围内现有湖水及河水的水质报告？

答：可提供水文地质勘探报告，但由于没有对现有湖水及河水做系统的水质检测，现状湖水是依靠机井抽取地下水补给，河道主要为排污河道，无法提供水质报告。本次设计中可不考虑现状水质问题。

问题2

能否提供森林公园中的水系的流量、流向、水源，该等水系的水是源自城市的供水系统，还是引入清河的水？可否告知现状洼里片林水域的正常水位？

答：森林公园中水系的水源在奥运赛事期间使用城市水系的清水，由京密引水渠昆玉河提供。清水经长河、转河、小月河及辛店村路拟建输水暗沟于北辰西路附近进入奥运湖，以常流水方式按1.0m³/s流量补充，赛前将湖水放空并注入清水，奥运赛事周期按30天考虑。平时使用城市中水，由白庙村路拟建中水管道提供，中水先流经湿地，再进入奥运湖，考虑蒸发、渗漏及换水量，经测算森林公园中水系的日均需水量约为5.74万m³，折算成流量为0.664m³/s。根据森林公园中的地势和水源补给方位，森林公园中水系的流向应是自西向东和自南向北。

洼里片林水域依靠每天抽取地下水进行补给，且湖底没有衬砌，渗漏较大，无法提供正常水位，设计当中可不考虑。

问题3

附图03 与附件01 中的设计范围不同，请确认应以哪个图为准？

答：以图03 为准。

SOLICITING FOR LANDSCAPING OF FOREST PARK AND CENTRAL ZONE IN OLYMPIC GREEN INQUIRY/CLARIFICATION ABOUT PROJECT PRESENTATION

The intelligent property of any information (including but not limited to written material, drawing, picture, photograph, and electronic file, etc.) provided for the Participants by the Organizer, especially the copyright, belongs to the Organizer or other copyright owners in question. Any Participant shall neither provide such information for any third party, nor use it for any purpose other than for preparing his landscaping of Forest Park and Central Zone in the Olympic Green.

Question 1:

Are there any available data on geological investigation in the Olympic Green and any report of present lake/river water quality in the plot area?

Answer:

Hydrogeological data are available, but without any water quality report as systematic water quality testing of existing lakes/rivers has not been conducted. The existing lakes are replenished with underground water pumped out by motor-pump. The rivers are mainly used for discharge. The water quality of lakes/rivers may be neglected in the consideration of design.

Question 2:

Are there any data on the flow, direction, and source of water systems in the Olympic Green? Are these water systems fed by the municipal water supply system or the Qinghe River? How high is the normal level of existing Walipianlin waters?

Answer:

The source of water systems in the Forest Park will be clear water from the municipal water system during the Olympic Games. Clear water will be introduced from Kunyu River and Jingmi Diversion Canal into the Olympic Lake at Beichenxi Road via the Changhe, Zhuanhe, and Xiaoyuehe Rivers, and the proposed water-conveyance blind ditch along Xindiancun Road, to replenish the Olympic Lake at a permanent flow rate of 1 m^3/s. The lake will be emptied and filled with clear water before the Olympic Games assumed to last for a period of 30 days. In other times, municipal intermediate water supplied by the proposed intermediate water pipes along Baimiaocun Road will be introduced into the lake after treated in wetlands. With evaporation, seepage, and replenishment taken into account, the daily water demand of such water systems is estimated to be about 57 400 m^3/d, equivalent to a flow of 0.664 m^3/s. Based on the terrain of the Forest Park and the direction of replenishing water source, the flow direction of water systems should be from west to east and from south to north.

The existing Walipianlin waters are replenished with underground water pumped out by motor-pump. Because the bottom is without liner and the leakage problem is serious, the normal level of existing Walipianlin waters is not available to be measured. The normal level of existing Walipianlin waters may be neglected in the consideration of design.

Question 3:

The scope of design shown in Figure 03 is different from that shown in Annex 01, which is correct?

Answer:

The scope of design shown in Figure 03 is correct.

问题 4
附图 07 "交通空间"、"楼梯空间"指什么，在图中的哪个位置？

答：附图 07 交通空间、楼梯空间在森林公园用地南侧奥林匹克中心区建筑群中，不在方案规划设计范围内。

问题 5
附图 09 与 03 中 B 区南端的后勤设施是否为临时建筑，赛后是否拆除？

答：B 区南端的后勤设施是临时设施，赛后拆除。

问题 6
附图 09 轴线北端蓝色区域是什么？

答：附图 09 轴线北端蓝色区域为意向中的公园主入口及主要活动区，在奥运会期间作为赛事组织使用空间。

问题 7
CAD 05 图（奥运支线平面图）中，公交站点、热力点在图例中有显示，但图中没有显示，它们的具体位置在哪里？

答：公交站点的位置应以有关的道路交通规划图为准，热力点的布置原则是以每个建筑地块为单位，按负荷需求新建自用热力点。

问题 8
大屯路下穿中心区，但图中显示未标明下穿的位置，可否告知具体的位置？

答：大屯路下穿中心区起点北辰西路向西 200m，终点北辰东路向东 200m，且地面亦有道路。

问题 9
可否详细说明成府路、大屯路、北四环路等交通道路与中轴线处的交叉关系（是地下穿越中轴线，还是被截断了）？可否提供北辰西路／五环路立交桥的平面图？

答：北四环路、成府路和大屯路与中轴线交叉时均下穿中轴线，其中大屯路、成府路除有地下道路外，地面有路与中轴线平面交叉，除上述三条路与中轴线立体交叉外，其余各交通道路均与中轴线平面交叉。

问题 10
北中轴路是跨越北四环路还是和景观广场在一个平面上？

答：北中轴路跨越北四环路，比现在地面高约 2.5m，向北逐渐与地面接平，和景观广场在一个平面上。

问题 11
北小河的位置在哪？

答：北小河位于北苑路以东，不在本项目的范围之内。

问题 12
请确认热身场在赛时及赛后的具体位置。

答：国家体育场热身赛场位置尚未最终确定，初步定于主场北侧、中轴线东侧，拟于第二次答疑时提供参考平面位置。

奥林匹克公园森林公园及中心区景观规划设计方案征集项目情况介绍会质疑／澄清

SOLICITING FOR LANDSCAPING OF FOREST PARK AND CENTRAL ZONE IN OLYMPIC GREEN INQUIRY/CLARIFICATION ABOUT PROJECT PRESENTATION

Question 4:
What does it mean by saying "traffic space" and "stair space" in Figure 07? Where are they shown in the figure?

Answer:
The traffic space and stair space shown in Figure 07 are in the building complex in the Central Zone on the south side of the plot area of the Forest Park, not in the scope of design.

Question 5:
Are the logistic facilities at the south end of Zone B shown in Figures 09 and 03 temporary ones to be removed after the Olympic Games?

Answer:
Yes, they are temporary facilities to be removed after the Olympic Games.

Question 6:
What is the blue space at the north end of the axis shown in Figure 09?

Answer:
The blue space at the north end of the axis shown in Figure 09 is identified for the proposed main entry and main activities. It will be used for event operation during the Olympic Games.

Question 7:
CAD05 (plan of Olympic Subway Branch) includes legends of bus stops and heating plants, which, however, are not shown in the drawing, where are their specific locations?

Answer:
Bus stops are located as shown in relevant traffic planning charts, while heating plants are built by users according to the demands of individual building lots.

Question 8:
Datun Road underpasses the Central Zone, but the underpass section is not shown in the drawing, where is the specific location?

Answer:
The underpass section in the Central Zone starts 200 m west of Beichenxi Road and ends 200 m east of Beichendong Road. This road also includes a ground section here.

Question 9:
Would you please detail Chengfu, Datun, and North Fourth Ring Road crossing with the central axis (do they underpass the central axis or are they cut off)? Is there any available plan of Beichenxi Road/Fifth Ring Road exchange?

Answer:
All of the North Fourth Ring Road, Chengfu Road, and Datun Road underpass the central axis. Datun and Chengfu Roads also have level crossings with the central axis. While these 3 roads have grade separation with the central axis, all other roads have level crossing.

Question 10:
Does Beizhongzhou Road overpass the North Fourth Ring Road or cross the landscape plaza?

Answer:
Beizhongzhou Road overpasses the North Fourth Ring Road at a height of about 2.5 m, gradually lowers to ground in the north, and crosses the landscape plaza.

Question 11:
Where is the Beixiaohe River?

Answer:
This river is located east of Beiyuan Road, beyond the project area.

Question 12:
Would you please confirm the specific location of the warm-up area during and after the Olympic Games?

Answer:
The warm-up area is not yet finalized, but it is tentatively located north of the National Stadium east of the central axis. It is proposed to include a plan for reference in the second written response.

Question 13:
Is the function of "reserved space" specified in the planning? When will construction be started in such "reserved space"?

Answer:
Such reserved space is reserved for future development of Central Zone after 2008(as shown in Figure 04). Block B24 and B26 was planned for the huge public facilities space as

奥林匹克公园森林公园及中心区景观规划设计方案征集项目情况介绍会质疑／澄清

问题 13
在规划中对"预留用地"的功能有无规定？该等"预留用地"预计什么时候开始建设？
答：附图04中注明的预留发展用地，是为奥林匹克公园中心区以后的长远发展预留用地。地块B25、地块B26规划为大型公共设施用地，地块B23、地块B24规划为商业开发用地。预留发展用地将会在奥运会以后选择合适时机建设。

问题 14
可否告知森林公园主入口之外的其他入口的位置以及停车场的位置？
答：建议入口数量按一般规范进行设计，安排在用地的四周。

问题 15
是否取消了轻轨？若未取消，轻轨是否包括在本次景观规划设计的范围内？
答：该项目尚需深入进行可行性论证，应征人可结合方案提出建议。

问题 16
请解释奥林匹克公园内各条道路的通车情况？
答：城市道路可参照已明确的设计规模进行推算，公园内部道路可参照一般规范，根据各应征人规划设计的功能、项目分布情况进行设计。

问题 17
地下市政管网是否采用综合市政管廊的形式？其平均埋深是多少？与地铁的关系如何？
答：奥林匹克公园及周边城市道路上的地下市政管网均采用直埋敷设方式。除交叉路口外，一般埋深在－8.0m以内。与地铁交叉时，市政管线均由地铁上方穿过。

问题 18
可否提供奥运村的总平面图？
答：目前奥运村的设计方案尚未最终确定，待近期确定后将作为补充材料提供。

问题 19
可否提供项目情况介绍会的PPT演示文件？若该PPT的演示文件与征集文件不同时以哪一个为准？
答：以征集文件为准。

问题 20
能否更加清晰地阐述由Sasaki设计的获选的奥林匹克森林公园的总体规划对下一阶段设计的指导作用？是否要以这个方案为基础进行设计？是否可以修改其设计理念甚至方案？还是要延用原方案？
答：SASAKI方案的一些概念应予以维持，如山水的概念、中轴线与森林公园的空间关系、中轴线西侧现状的建筑空间与东侧自由灵活水系的对比关系等，在此基础上，可结合方案进行设计。

问题 21
对原奥林匹克公园总体规划"龙形景观"、"千年文明轴"的尊重程度应如何把握？水体位置的可行性？与现有森林的矛盾？森林公园的主水体可以规划在别的位置吗？

well as B23 and B24 was planned for commercial development. Such reserved space will be constructed at some proper time after the Olympic Games.

Question 14:
Where are other entries and parking areas located except for the main entry?

Answer:

It is recommended to consider the number of entries as per general specifications, and arrange them around the plot area.

Question 15:
Is the trolley cancelled? If not, is it included in the scope of design?

Answer:

The feasibility of this aspect is to be justified in depth, but the Participant may provide recommendations in terms of his scheme.

Question 16:
Would your please give the traffic volume of roads in the Olympic Green?

Answer:

The traffic volume of urban roads is estimated on the basis of the determined design scale, while that of internal roads in the Olympic Green is considered, by referring to general specifications, on the basis of the function and distribution of the facilities envisaged by the Participant.

Question 17:
Are underground municipal pipelines arranged as integrated municipal pipe galleries? What is the average depth? How are they related to the subway?

Answer:

Municipal pipelines in the Olympic Green and along the urban roads around are all directly laid, generally at a depth of -8 m except at road crossings. They all overpass the subway where there is a crossing.

Question 18:
Is there any available general layout of the Olympic Village?

Answer:

Such will be provided as additional information when the Olympic Village is finalized soon.

Question 19:
Will you provide a PPT file of project presentation? If there is any discrepancy between such PPT file and the Soliciting Documents, which will prevail?

Answer:

The Soliciting Documents will prevail.

Question 20:
Would you please describe more clearly how the general planning of the Forest Park developed by Sasaki will guide the next stage of design? Will the design be based on such scheme? Can the design concept and even the scheme per se be changed, or will the original scheme be continued?

Answer:

Some concepts included in the scheme of Sasaki will be maintained, such as the concept of hills and waters, the spatial relationship between the central axis and the Forest Park, and the contrast between the existing building space on the west side of the central axis and the free water systems on the west side. On this basis, the design is to be developed in light of the said scheme.

Question 21:
How to respect the "dragon-shape landscape" and "axis of millennium civilization" proposed in the original general planning of the Olympic Green? Is the location of waters feasible? Is there any conflict with the existing forest? Can main waters in the Forest Park be positioned elsewhere?

Answer:

The design is to be developed in accordance with the written materials and drawings contained in the Soliciting Documents, with landscape along the central axis designed to establish a landscape series and create space for human activities. The theme of the central axis is not specified in the Soliciting Documents, so recommendations can be offered in light of the

答：请根据征集文件提供的文字、图纸等资料进行设计，其中，中轴线的景观设计要考虑形成一定的景观序列及人的活动空间等，关于轴线的主题征集文件未予限定，可结合方案提出建议。水体规划应充分考虑对现状成形片林的保护，但水面的整体形状和位置不宜作较大改动。

问题 22

可否对横穿森林公园的北五环路段进行改造？比如将部分路段变成地下路段？

答：不建议对北五环路段进行改造，除非有非常特殊的必要。

问题 23

森林公园内现有高架桥横跨五环路连接五环路两侧的公园内的两个地块，该高架桥是否必须保留？

答：该高架桥不属于城市规划道路市政工程，可考虑拆除、改造、利用等设计方案。

问题 24

对于建筑基地、地铁、水体挖出的土方量，是否应在奥林匹克公园内就地平衡？

答：提倡森林公园内的土方平衡，如无法满足，可考虑利用场馆区的土方予以补充。

问题 25

征集文件中的"服务设施占全园陆地面积的比例控制在建设用地的3％以内"是指建筑基底？还是建筑占地？还是建筑面积？

答：3％指的是建筑占地面积，御苑花园别墅的建设占地已计算在该指标中。

问题 26

征集文件中要求水面面积100hm^2。但现有水面仅12hm^2，增加如此多的水面，水源如何解决？这样的要求是否实际？可否适当缩小水面？森林公园中湿地和水的面积可否调整，调整的余地可以有多大？

答：水源的问题已在问题1中澄清。森林公园中水面的形状、走向和面积不宜作较大改变。湿地面积应符合征集文件中的要求。

问题 27

请详细告知多媒体演示文件中的配音解释、背景音乐及三维动画等方面的要求？

答：应征人提供的多媒体演示电子文档可使用VCD、DVD或PPT格式。演示时间在5分钟以内，演示文件可配以背景音乐和中、英文字幕，但不应配以语音解说。

奥林匹克公园森林公园及中心区景观规划设计方案征集项目情况介绍会质疑／澄清

SOLICITING FOR LANDSCAPING OF FOREST PARK AND CENTRAL ZONE IN OLYMPIC GREEN INQUIRY/CLARIFICATION ABOUT PROJECT PRESENTATION

scheme. The planning of waters should include due attention to the protection of established continuous forests, but it is inadvisable to make any major change to the overall shape and location of waters.

Question 22:
Is it possible to reconstruction the section of the Fifth Ring Road crossing the Forest Park, for example to change it into a underpass section?
Answer:
Reconstruction is not recommended unless with special necessity.

Question 23:
Is it necessary to maintain the existing viaduct that saddles the Fifth Ring Road and bridges the land parcels on both sides of this road?
Answer:
This viaduct is not in the category of planned urban roads. Removal, reconstruction and other arrangements can be considered.

Question 24:
Is it necessary to balance the soil excavation volumes of building foundation, subway, and waters in the Olympic Green?
Answer:
Balancing soil excavation volumes within the Forest Park is encouraged. If there is any failure to do this, however, soil materials from other parts of the Olympic Green may be considered as additional sources.

Question 25:
It is stated in the Soliciting Documents that "The percentage of these facilities in the whole land area shall be controlled within 3% of the plot area..." does it mean the building foundation, building area, or floor area?
Answer:
This 3% means the percentage of the building area. Also, it is noted that Yuyuan Garden Villa is included.

Question 26:
In the Soliciting Documents, it is required to have a water surface of 100 hm^2, but there exists 12 hm^2 only. How to feed such an additional water surface? Is this a practical requirement? Can the water surface be reduced as appropriate? Can the surface of wetlands and waters in the Forest Park be adjusted? If yes, how large is the room of adjustment?
Answer:
The source of water is clarified in Question 1 above. It is inadvisable to make any major change to the shape, direction, and surface of waters in the Forest Park. The surface of wetlands shall be as required in the Soliciting Documents.

Question 27:
Are there any specific requirements for the dub, background music, and 3-D cartoon of multimedia files?
Answer:
The multimedia files submitted by the Participant may be prepared in VCD, DVD or PPT format, to demonstrate for a period of 5 minutes. Such files may be accompanied by background music and Chinese/English captions, but without voice commentary.

AOI
EDAW Inc. 易道公司

China Architecture Design & Research Group
中国建筑设计研究院

奥林匹克公园森林公园及中心区景观规划设计方案

DESIGN SCHEMES FOR LANDSCAPING OF FOREST PARK AND CENTRAL ZONE IN OLYMPIC GREEN

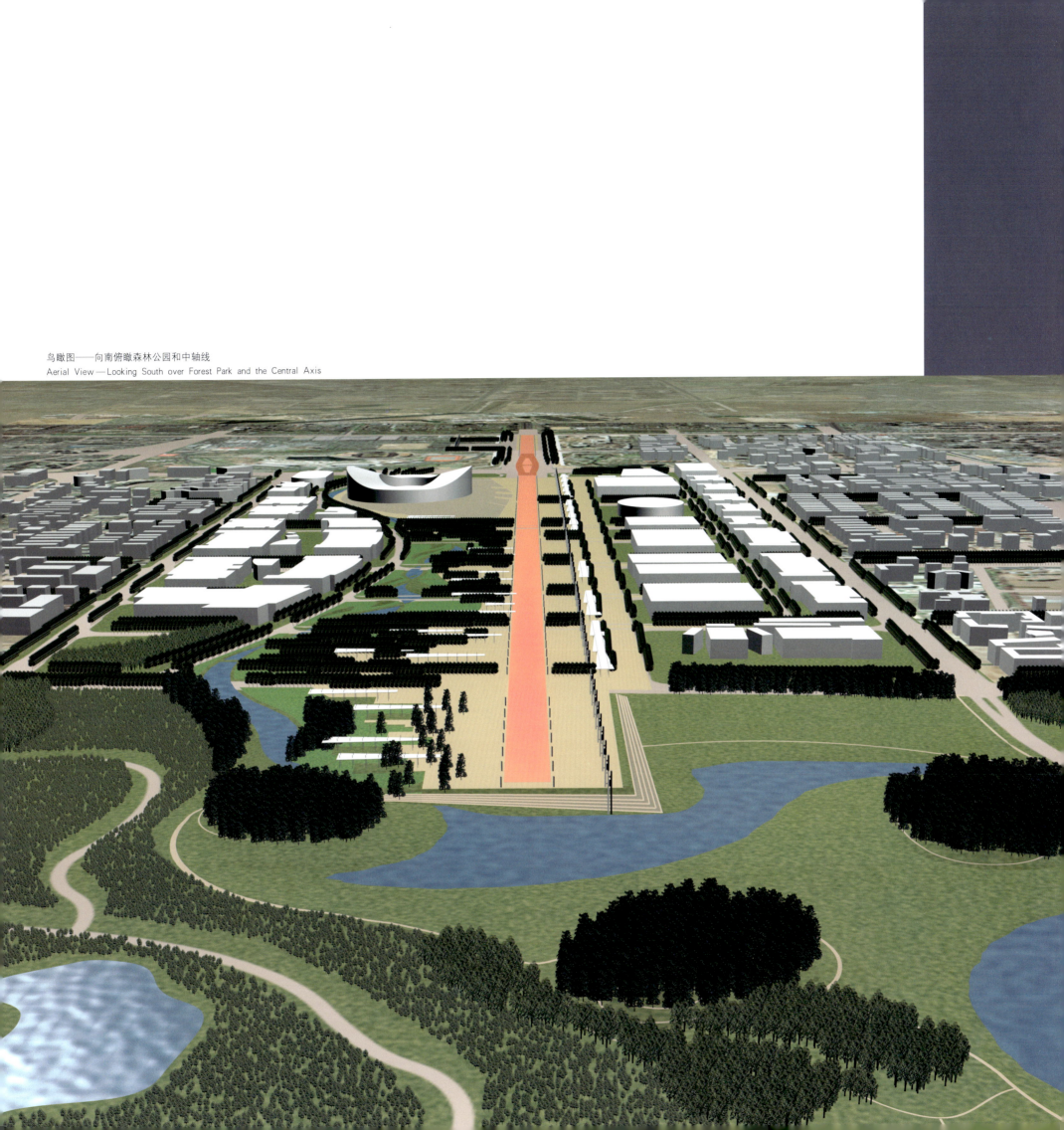

鸟瞰图——向南俯瞰森林公园和中轴线
Aerial View — Looking South over Forest Park and the Central Axis

鸟瞰图——向西北俯瞰奥林匹克中心区和森林公园
Aerial View — Looking toward the Northwest over the Central Olympic Zone and Forest Park

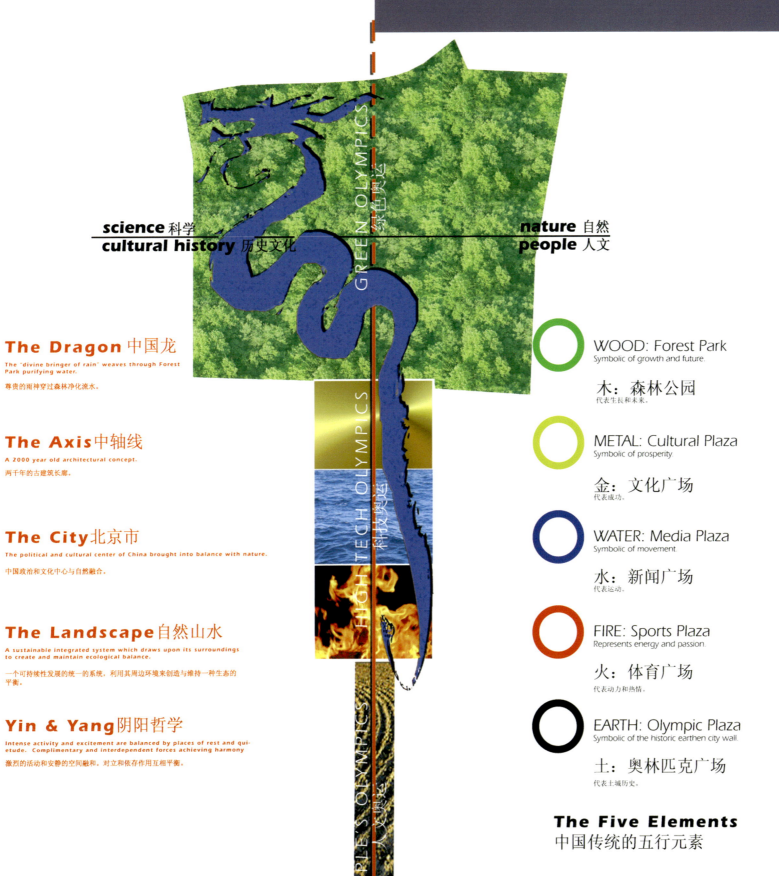

science 科学　　　　　　　　　　　　nature 自然
cultural history 历史文化　　　　　　people 人文

The Dragon 中国龙
The "divine bringer of rain" weaves through Forest Park purifying water.
尊贵的雨神穿过森林净化流水。

The Axis 中轴线
A 2000 year old architectural concept.
两千年的古建筑长廊。

The City 北京市
The political and cultural center of China brought into balance with nature.
中国政治和文化中心与自然融合。

The Landscape 自然山水
A sustainable integrated system which draws upon its surroundings to create and maintain ecological balance.
一个可持续性发展的统一的系统，利用其周边环境来创造与维持一种生态的平衡。

Yin & Yang 阴阳哲学
Intense activity and excitement are balanced by places of rest and quietude. Complimentary and interdependent forces achieving harmony
激烈的活动和安静的空间融和。对立和依存作用互相平衡。

○ WOOD: Forest Park
Symbolic of growth and future.
木：森林公园
代表生长和未来。

○ METAL: Cultural Plaza
Symbolic of prosperity.
金：文化广场
代表成功。

○ WATER: Media Plaza
Symbolic of movement.
水：新闻广场
代表运动。

○ FIRE: Sports Plaza
Represents energy and passion.
火：体育广场
代表动力和热情。

○ EARTH: Olympic Plaza
Symbolic of the historic earthen city wall.
土：奥林匹克广场
代表土城历史。

The Five Elements
中国传统的五行元素

DESIGN SCHEMES FOR LANDSCAPING OF FOREST PARK AND CENTRAL ZONE IN OLYMPIC GREEN

连绵分布的植被
Successional Plantings

可持续性水体系统
Sustainable Water System

可持续性植被
Sustainable Plantings

本方案的景观设计本着两个中心,一个是短期的,即奥运会期间的运营成功,另一个是长期的可持续性使用,即赛后奥运遗产的运营。设计将对奥运赛后遗产的利用模式进行全面的设计。设计旨在建立一个以富有活力的公共场所为中心的景观,其中包含了富有吸引力的多个景点。此外,设计采用了中国园林的可持续性的长期发展的设计理念。设计框架本身具有很大的灵活性,在广场和公园地区可以增加新的活动,这保证了设计的长期可持续性。将来需要时,可以增加商业设施、文化设施或旅游设施。

The design of the landscape is focused both on the short term, Olympic success and the long term sustainability, legacy agenda. Our design will have integrity during the Games and longterm sustainability. In the legacy mode the full design will be implemented. Our design aims to establish a landscape with a focus on vibrant public places within which a variety of attractions will be located. In addition our design adopts a long term sustainable planting strategy for the gardens. Our long term sustainability is ensured by the inherent flexibility of the design framework to accept the addition of new activities within the plazas and park areas. These additions could be more commercial facilities, more cultural facilities or more tourist facilities as may be needed in the future.

奥运赛后经营模式 人文、生态的长期可持续性
Legacy Mode cultural & biological long term sustainability

奥运期间运作模式 戏剧性、丰富的体验、与自然的和谐统一
Olympic Mode & drama, richness of experience and harmony with nature

DESIGN CONCEPT
设计理念

1. 中国龙

在本方案的设计中,水是最为重要的元素。在中国的神话与传统文化中,龙是最为尊贵与神圣的图腾,总是与水结合在一起,被视为可以呼风唤雨,集天地之精华,象征着富贵与吉祥。在风水学中,龙被认为是地之气,龙脉行走在山的下面,是看不见而有生气的东西,如人之神经,土是龙的肉,石是龙之骨,草木是龙之毛发。郭璞在《葬书》中曾言:"气乘风则散,界水则止。"因而,古代的造园师、工匠、风水师,便常通过水来收集与聚集气,阻止其散开。在这个思想的引导下,龙的形象被引进了中国古典园林设计的概念之中。龙在其中是一种神秘的象征,常常通过水的聚集与净化而现身。

在奥林匹克森林公园与中心区的设计中,我们引入了"龙"的概念。龙的身体蜿蜒穿越基地,张开的龙嘴对着清河,而其尾巴则环绕着体育馆。龙嘴正是水引入基地、水处理开始的地方,龙的尾巴则是奥林匹克大门所在之处。奥林匹克大门象征着中国过去的成就与未来的理想之结合点,水自喷泉中冲出,进入空气之中,与龙的概念相结合,庆赞着奥林匹克盛会与北京古城的新世纪。

将中国文化一贯崇尚的龙图腾引入景观设计之中,使其成为北京与中国新纪元的象征。即,人、科技、自然共同合作,创造一个健康与可持续性发展的环境。基于龙元素的设计,将雨、水元素与当代的环境结合起来,来建设一个具备可持续性潜力的水景,不仅能够四季常新,而且实现了地下水供应的自然更新。

在设计中,龙的象征也体现在地形之中。九个突起的堆土节点,包括奥林匹克体育馆,穿过森林公园与中心区,依次排开。这九个堆土节点的设计灵感,则来自于"龙生九子"的故事。传说龙生九子,而各有所好。这九个堆土节点也因此具有独特的个性特征。关于这个传说的记载各不相同,比较常见的说法是:

一曰嘲风:喜好冒险,因而人铸其像,置于殿角;
二曰睚眦:平生好杀,喜血腥之气,其形为刀柄上所刻之兽像;
三曰鸱吻,最喜欢四处眺望,常饰于屋檐上。
四曰趴蝮,性喜水,常饰于石桥栏杆顶端。
五曰蒲牢:平生好鸣,它的头像被用作大钟的钟纽;
六曰狴犴,平生好讼,形似虎,狱门或官衙正堂两侧有其像;
七曰囚牛:性喜音乐,其形为胡琴琴杆上端的刻像;
八曰狻猊,性好烟火,常饰于香炉盖子的盖钮上。
九曰椒图,形状似螺蚌,性好闭,常饰于大门口,铺首衔环是其形象。

1. The Dragon

Water is an essential aspect of our design. In Chinese mythology, there is a dragon--Lung--that is regarded as the "divine bringer of the rain". Lung is necessary for human health and well-being. Dragon is Qi of Earth. The classics say: 'When Qi rides the wind, it is scattered; when it encounters water, it is retained.' Thus, the ancient masters gathered and collected Qi through water to prevent it from being scattered. Inspired by these notions, the form of a dragon is etched into the landscape by means of water and earth. He is a mysterious creature and reveals himself periodically through the action of water collection and purification.

The Dragon's mouth is open to the Qinghe River, his body winds through the site, and his tail wraps around the Olympic Stadium. His mouth marks the beginning of the treatment process for water entering the site and his tail delineates the Olympic Gateway. The Gateway is the juncture of previous achievements and future ideals for China and the world. Water jets stream into the air at this point in celebration of both the Olympics and Beijing.

The meaning that the dragon embodies is inscribed in the landscape as a symbol of a new era for Beijing and for China. This is an era where man, technology, and nature work together for the creation of healthy and sustainable environments. The designed embodiment of the dragon combines the phenomena of a rain storm with the environmental technology of today to make a sustainable waterscape that can replenish seasonally and recharge the groundwater supply.

Sketch View from Plaza toward Golden Bridge
从广场望向金桥

Olympic Green

The symbolism of the dragon is also reflected in the landforms created in the design. The nine mounds, including the Olympic Stadium, are spread through the Forest Park and Central Zone. These mounds are inspired by the nine sons of the Dragon. Each son had a strong personality as does each of the mounds. There is no general agreement as to what the sons are called. However, to most people, they are:

(1) Haoxian - A reckless and adventurous dragon whose image can be found decorating the eaves of palaces.
(2) Yazi - Valiant and bellicose; his image is seen on sword-hilts and knife hilts.
(3) Chiwen - Chiwen likes to gaze into the distance and his appearance is often carved on pinnacles.
(4) Baxia - Baxia is a good swimmer and his image decorates many bridge piers and archways.
(5) Pulao - Pulao is fond of roaring and his figure is carved on bells.
(6) Bixi - Bixi is an excellent pack-animal whose image appears on panniers.
(7) Qiuniu - Qiuniu loves music and his figure is a common decoration on the bridge of stringed musical instruments.
(8) Suanni - Suanmi is fond of smoke and fire; his likeness can be seen on the legs of incense-burners.
(9) Jiaotu - Jiaotu is as tight-lipped as a mussel or a snail. His image is often carved on doors.

Sketch View from Water Theater
水剧场透视图

2. 阴阳哲学

在中国博大深厚的文化中，阴阳哲学是其中一颗璀璨的明珠。它体现了古人对事物的辨证思维，其内涵蕴涵了宇宙世界的至理，体现了自然运行的本质规律。老子云："天下之物，有生于无，道生一，一生二，二生三，三生万物。"古人认为，阴阳乃为万物之本。一阴一阳的对立存在不是静止的，而是运动的，以阴阳形式存在，负阴而抱阳，对立存在既为阴阳，以树叶为例，正面朝上的为阳，背面为阴，以人为例，面为阳，背为阴，以人类性别论，男为阳，女为阴，以天地论，天为阳，地为阴。以此观之，宇宙万物皆有阴阳，阴阳无处不在，小可至微，大可宇宙，一套阴阳理论解释了万物无不由阴阳交合而存在。阴与阳不会单独存在，负阴而抱阳明确指出，阴中有阳，阳中有阴，万物皆然。

根据阴阳哲学，自然界被看作是一个不断变化的具有生命的机体。当其运动之时，其产生的能量为阳，当其静止之时，产生的能量为阴。阴阳代表着宇宙万物之间的平衡与互动。同样，将阴阳哲学融入到设计之中，便意味着各个设计元素之间的平衡，包括坡度、形状、景观风格以及位置等元素，使不同的景观之间相互融合，相互支持。奥林匹克中心区，以其密集的活动、高昂的氛围，代表着设计中的"阳"。而森林公园，以其安静、悠闲、低密集的人群，代表着设计中的"阴"。

阴阳概念加入到基地的设计之中后，阴阳能量将互相衬托、互相糅合，形成和谐的平衡。低速的水流系统与净化系统，包括30hm²的人工湿地、横向／竖向的芦苇池，作为活动能量低且缓慢的区域，成为"阴"的象征；较为活跃的水流系统，诸如流水、人工水底换气以及喷泉等，则成为"阳"的象征。在平面图上，可以看到具有季节性变化的生态环境（阳）与稳定而较少变化的植栽区（阴）之间的对比与共存。绿色缓冲带、地形以及硬景用于隔离活动区与其他区，或者减少"负面"能量（如高速路）的影响。绿色能量系统也被分为被动的阴（如太阳能）与主动的阳（如风能、燃料）。

Olympic Green

2. Yin and Yang

Yin and Yang, the two primordial forces that govern the universe, symbolize harmony. They are opposites: Yin is dark, Yang is light, Yin is passive, Yang is active, Yin is female, Yang is male. Yin and Yang are complementary, they depend on each other - they interact continuously and produce all forms of existence on earth.

Nature is regarded as a living organism that breathes unceasingly. When it moves, its breath produces Yang energy.

When it rests, its breath produces Yin energy. The Yin and Yang are important for the balanced, mutual operation of the whole. In this same way we aim for a balance of design features (slope, forms, landscape patterns, and placement) that are different in form but placed to balance and support each other.

Our design incorporates Yang energy exhibited by places of intense activity and excitement in the Central Zone, and Yin energy exhibited by places of rest, quietude, and escape from the crowds in Forest Park.

This concept translates to the design of well-drained slopes where Yin and Yang energies are transpositioned and to slope arrangements that support our mandate for creating passive treatment (Yin) for slow water circulation and purification (including 30 hectares of constructed wetland, horizontal and/or vertical flow reed beds), but also potentially including more active water systems (e.g., flow forms, bubbler aerators, fountains, etc.) that embody Yang energy. Our design articulates juxtapositions of forest and other cover types to lay out areas of seasonally changing ecological dynamics (Yang) with areas of constancy and slow change in the vegetative cover (Yin). Vegetated buffers, landforms and hardscape features are used to separate activity areas and block or isolate "negative" energies (e.g., highway traffic). Green energy systems are considered and represent passive Yin (e.g. solar) and active Yang (e.g. wind, fuel cell) energy development.

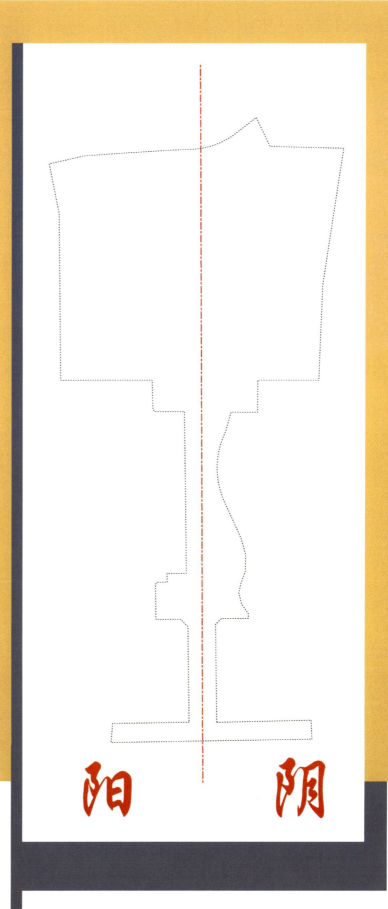

Olympic

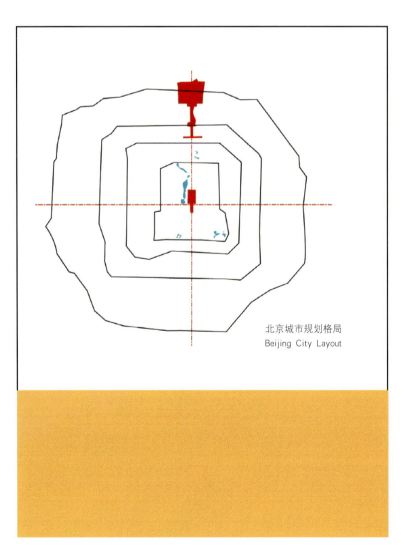

北京城市规划格局
Beijing City Layout

3. 中轴线——两千年的古建筑长廊

众所周知,紫禁城,是北京古都的核心所在,整个城市以其为中心点向外扩张。作为明清以来的历代皇宫所在,紫禁城以其尊贵、宏大、壮观与华丽而闻名于世。紫禁城的设计,体现了中国传统的儒家思想的核心:"中正"与"仁和"。今天,虽然紫禁城仍然以其巍然不动的高贵姿态位于21世纪国际性大都市的中心点,代表着过去与历史,辉煌与沧桑,而在北京的外缘带,却在悄然诞生着新一代城市标志性空间。

800多年来,北京一直是中国的政治与文化中心。北京位于华北平原的北端,2000多年来,城市所在的位置一直是连接中国东北、蒙古以及中国中部的贸易之路的交汇点。公元1272年,元朝皇帝忽必烈在北京建大都城,北京第一次成为统一国家的首都,开创了一代繁华盛景,元朝成为世界上最强盛的国家。而在此之前,北京也曾作为不少朝代的首都与中心城市。因而,北京拥有着从13世纪以来丰富的中国城市变迁的历史积累,而这也将影响现代的北京城市重塑工程。

忽必烈迁都燕京(即今北京)之后,将其改名为大都,并邀请出家还俗的学者刘秉忠主持元大都的设计,此人通《易经》,精阴阳。大都的设计完全恪守春秋时期的《周礼·考工记》中的布局。

《考工记》中记载:"匠人营国,方九里,旁三门,国中九经九纬,经涂九轨,左祖右社,面朝后市。"元大都正是根据该思想而建立的,全城以南北为中轴,居民四合院左右对称排列,宫殿的大明殿、延春阁等主题建筑均分布在中轴线上,呈一个较规则的长方形。今天,北京城的古代皇朝建筑在中国的多个古都中保留得最为完好。尽管在过去的几十年中,北京城已经发生了巨大的变化,无数的现代摩天高楼正崛起在这古老的土地之上,但北京内城仍然是古都的核心所在。故宫博物馆,即紫禁城,巍然而庄严的固守城中,两翼分别由不同的标志性空间包围着:右侧是中山公园,原是明、清时的社稷坛,是明清皇帝祭土地和五谷之神的地方,建于明永乐十九年(1421年);左侧是劳动人民文化宫,曾是明、清两代皇室家庙,旧称太庙,是封建王朝皇室供奉祖宗牌位、年节大典祭祀先人的地方,是现存最完整的明代建筑群之一。而紫禁城前后则是北京的城市南北中轴线之所在。

绿色奥运的设计,将继承和延续北京传统中轴线的风格,同时通过各建筑单体造型的新颖细致的设计,体现21世纪"新北京"的城市风貌。绿色奥运社区的建立,将决定北京城北部城区的未来城市功能、形象以及生活基调。绿色奥运的设计远景与目标已经确立。而我们的设计,将帮助进一步完善这个远景目标,并设定一个设计框架,使得未来的森林公园与中心区不仅能够继承与延续北京中轴线传统,而且能够体现21世纪百年古都的新风貌。

Green

3. The Central Axis——A 2000-Year-Old Architectural Concept

Beijing is known for the fact that at its center lies the Forbidden City, the great imperial palace of the Ming and Qing Dynasties. This remarkable artifact encapsulates the five Confucian virtues of humanity: sense of duty, wisdom, reliability and ceremonial propriety. While the Forbidden City will forever be embedded in the center of Beijing, the city is currently in a remarkable moment of rebuilding itself. The center of this remarkable city represents its history, however its edges are where its new identity will be made.

Beijing is the political and cultural center of China and has been so for more than eight hundreds years. Situated at the northern end of the North China Plain, the city was established at the intersection of trade routes connecting northeast China, Mongolia, and central China about two thousand years ago. It was the site of several dynastic capitals before the establishment of the Yuan Dynasty by Kubilai Khan in 1271. As such, the city encapsulates a historical record of Chinese urban philosophy dating from the thirteenth century and one that continues to influence contemporary city making in Beijing.

Originally called Dadu, it was a completely designed city, which adhered strictly to a traditional Chinese philosophy of Capital City construction. When the Yuan Dynasty decided to move its capital to Beijing, Liu Bingzhong, an accomplished scholar of the time, was made the chief architect. It was he who borrowed the architectural concept from the book Kao Gong Ji, written more than 2000 years ago.

In the book, the unknown compiler wrote that "in designing a capital, the architect lays it out nine by nine li (4.5 kilometers) with nine streets and avenues, three gates on each side, the ancestral temple on the left and an altar on the right of the palace, office buildings in front of the palace and a marketplace behind it ." Old Beijing was laid out exactly in accordance with that plan and is the best preserved among the ancient capitals of China. Despite the modern high rise buildings that have gone up in the past few decades, the inner city of Beijing still represents the ancient concept of the capital plan. The Palace Museum, which was the imperial palace, is flanked by two parks: on its right the Zhongshan Park, formerly the imperial altar, and on its left the Working People's Cultural Palace, formerly the imperial ancestral temple. Behind and in front of the former Forbidden City runs a straight street serving as the central axis.

The Olympic Green is the northern extension of the Central Axis of Beijing. Its creation will define the northern extension of the Axis and create a new area in the city of Beijing. Its design will inform how the northern part of the city functions, its image and the way that people will live for generations. There already exists a vision for what the Olympic Green will be. Our task is to help improve that vision and set a framework that can deliver to the people and city of Beijing a place which not only continues the urban design legacy of the Central Axis but also expresses the hopes and aspirations of Beijing as a 21st Century metropolis.

The Central Axis is expressed as five areas within our design scheme. Each of the five "plazas" has an overlay theme related to the five Chinese elements: metal, wood, water, fire and earth (soil). Each of plaza areas relates to one of the elements, which becomes the programmatic theme of the place. The five colors of the Olympic rings are an additional overlay related to the elements and expressed in paving, floral display, and furniture, poles, banners etc. The colors of the rings represent the continents. The five elements will be illustrated on the light towers and in dramatic video clips, which will literally be celebrations of the elements.

The Central Axis and the overlays of the elements culminates within the Forest Park in a celebration of the elements combined in nature.

4.中国传统的五行元素

　　中国传统的五行元素体现了自然万物的循环变化，这种循环变化表述为"阴-阳"的自然运动，这种能量运动就像五行元素——金、木、水、火、土本身的特性一样，它们各自并非以实际的形式存在，而是表达了在自然轮回中以不同的形式不断转化，在不断的从无极到太极的变换中相互作用彼此抵消，无始无终的循环。

　　在五行元素的系统中存在着一系列的相互的牵制与平衡作用。这些作用体现为"发"与"制"的循环出现。这种循环保持了五行元素的极性从而能够不断运动和转化。而五行元素的存在又使得循环运动不断进行一种元素抑制前一种元素而发生发展，再孕育产生了另一种元素而被抑制。这样进行无休止的循环往复而同时保持了整个系统的平衡。

　　与阴阳平衡非常相似，在我们的方案中因为结合了中国传统的五行元素而更增添了整体的平衡与和谐。同样道理，元素之间的相互作用表现为森林公园中自然系统的延续性与季节的变换轮回。

DESIGN
CONCEPT

THE FIVE ELEMENTS

4 .The Five Elements

The Chinese five elements represent the cyclical changes of nature which regulate life on earth. They are the physical manifestations of the energies of Yin and Yang. These energies or movements resemble the function and character of their respective elements: wood, fire, earth, metal, and water. The elements are not literally forms or substances, but represent various forms of transformation in the natural cycle. They interact and counteract one another in constant transformation of emptiness to abundance and abundance to emptiness. They act as a continuous ring with no beginning and no end.

Built into the Chinese system of the five elements is a series of checks and balances. These checks and balances are referred to as 'creative' and 'control cycles'. The cycles maintain the polarity between the elements, which is required to move and transform energies. Thus, it is the presence of the five elements that keeps the cycles moving, as one element overtakes or nurtures the other. A rhythmic sequence is established in this way with the ultimate effort always being to restore harmony to the system.

Much like the balance of Yin and Yang, the incorporation of the five elements in our design additionally contributes to an overall balance and harmony in the plan. Likewise, the cyclical rhythm of change created by the interaction of the elements reflects the attributes of seasonality and succession found in the natural systems of the Forest Park.

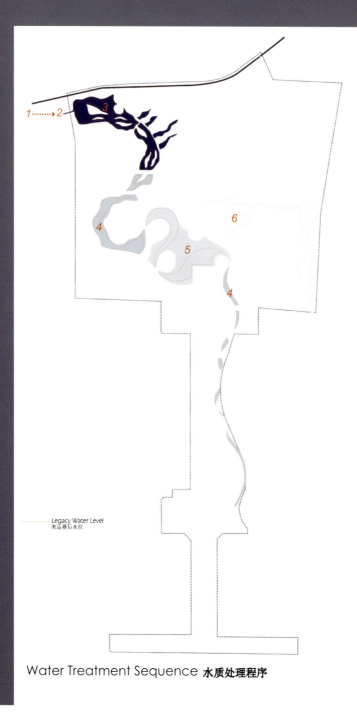

Water Treatment Sequence 水质处理程序

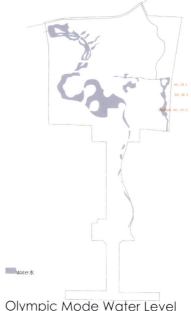

Olympic Mode Water Level
奥运期间水位

Legacy Mode Water Level
奥运赛后水位

Lagoon: Water is channeled from Qinghe River and pumped from the Qinghe Wastewater Treatment Plant into the mouth of the dragon (See Technical Indices for quantity and quality of water). The oxidation lagoon converts ammonia (NH_3-N) to nitrate (NO_3) and reduces the biological oxygen demand (BOD) of the water entering the system.

Wetland: Water is symbolically ingested by the dragon, filtered and cleaned as it passes through the body of the dragon. The water passes from the lagoon through a series of wetland islands and lakes where nutrient and pollutant removal takes place.

Lake: The chain of lakes remains throughout the year, while the full form of the dragon and the Olympic Lake appear only on rare and special occasions. Distinguished features of the dragon, such as the horns, are represented by shallow depressions in the land that act for flood control and recharge in higher volume storms, such as the 20 year design storm. On rare occasions, the full form appears and corresponds volumetrically to the 50 year design storm. This full form will occur during the Olympic Games.

氧化处理池：水由清河引入，然后经清河污水处理厂的专业化处理注入"龙"嘴之中（见水量与水质技术参数）。氧化处理池的功能是将氨水转化为硝酸盐（NO_3），并降低进入系统的水的生物氧（BOD）需求量。

湿地：流经处理池后，水流就像被"龙"吞进腹中，在流经龙的身体时到过滤和净化。从物理角度讲，从处理池流出的水经过一系列湿地，岛屿和湖泊，并在这些地方去除了水中的富营养物质和污染性物质。

湖泊：因此这一系列湖泊可全年存在，然而完整的中国"龙"和奥林匹克湖却只在极少的特殊情况下才能出现。"龙"的一些显著特征，例如角，可由地面的注流来表现；这些注地起着防洪作用，并会在遇到比较大降水量（例如20年一遇的大雨）时被水填满。在极少情况下，完整的"龙"才会出现，因为这相应地需要50年一遇的暴雨的水量才可实现。不过，在2008年奥运会时，为保证庆典视觉效果，整条龙都会出现。

Forest Park Storm Water & Waste Water Treatment & Reuse 森林公园雨水和污水处理再用

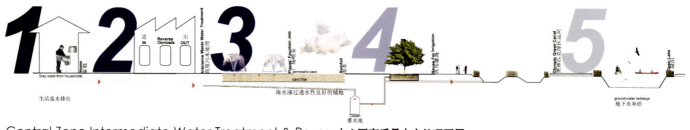

Central Zone Intermediate Water Treatment & Reuse 中心区高质量中水处理再用

水系概念图
Water Systems

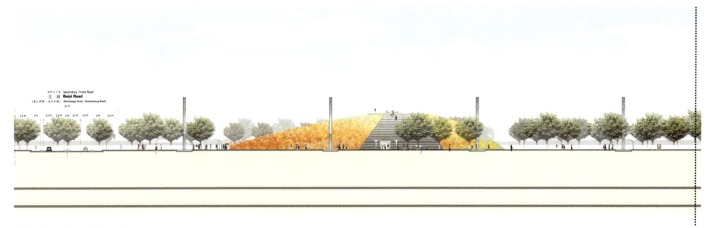

Beiyi Road to Bei'er Road
北二路到北一路

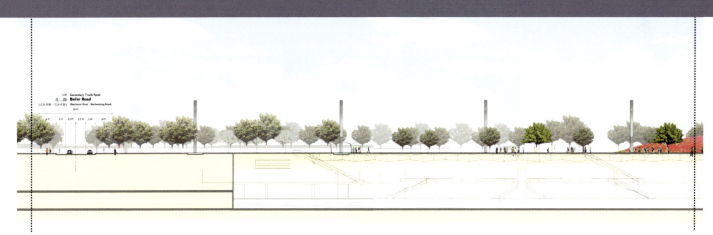

Bei'er Road and Through the Cultural Plaza
北二路和文化广场

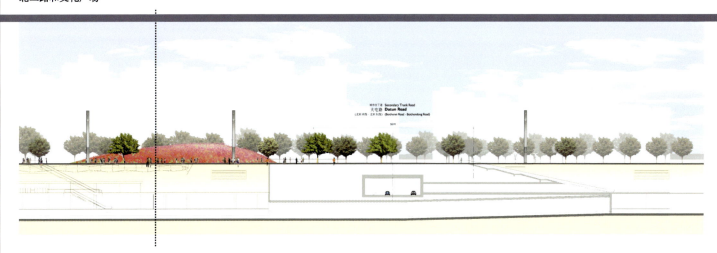

Through the Cultural Plaza to Datun Road
从文化广场到大屯路

基地剖面图 北京市的中轴线
Site Section the central axis, the line of Beijing

Sketch View From Plaza Toward Golden Bridge
从广场望向金桥

Central Zone of the Olympic Green
奥林匹克中心区

主要节点

中心区包含了以下几个重要的空间节点：本方案中的中轴线由五个广场区域组成。每一个"广场"都叠加了中国传统的"五行"元素，金、木、水、火、土。而每个广场都用其中一个元素来作为该广场的中心主题。除此之外，奥运五环中的五种颜色也被巧妙地与"五行"元素联系起来而以铺地、花木、城市家具、灯柱、旗杆等方式表达出来。众所周知，奥运五环的颜色代表五大洲。而相关的画面也将在沿中央轴线的灯塔上由巨型显示屏播放出来，从而吻合奥运五环的概念。

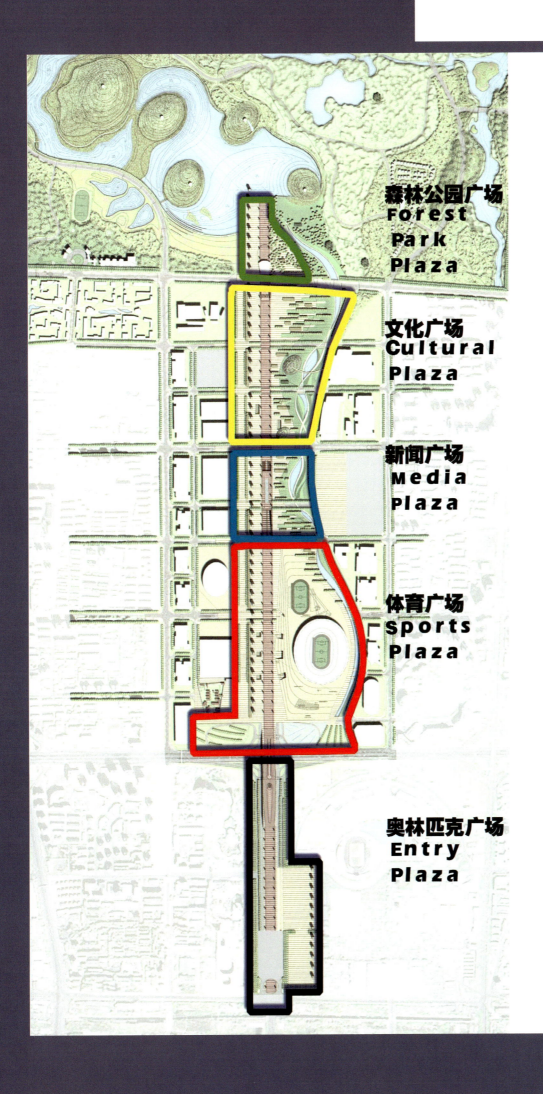

Major Nodes

The Central Zone consists of several nodes of activity arrayed along its length. These nodes are organized according to the five elements: earth, fire, water, metal, and wood. Each node consists of a plaza relating to one of the elements, which informs the programmatic theme of the place. Each element is additionally related to one of the five colors of the Olympic rings: black, red, blue, yellow, and green.

In Olympic tradition, these colors represent the five continents: Africa, the Americas, Europe, Asia, and Australia. Our design additionally incorporates the five colors and continents of the Olympic games by altering the color of paving, floral display, and site furnishings (i.e. benches, poles, banners) of each node. The five elements will be imprinted on the light towers and illustrated by dramatic video clips in celebration of the elements. During the Olympics, video clips will also be shown of each of the five continents participating in the games.

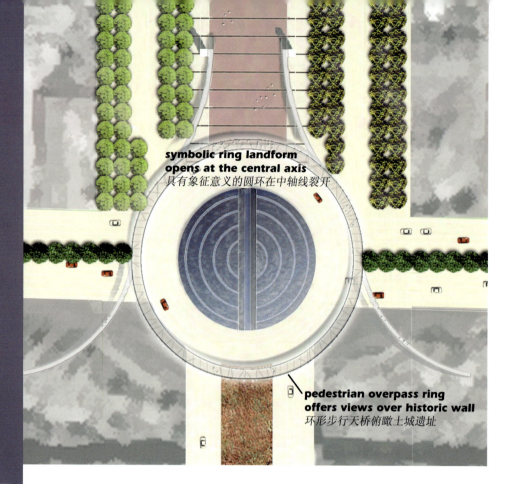

symbolic ring landform opens at the central axis
具有象征意义的圆环在中轴线裂开

pedestrian overpass ring offers views over historic wall
环形步行天桥俯瞰土城遗址

Plan
平面图

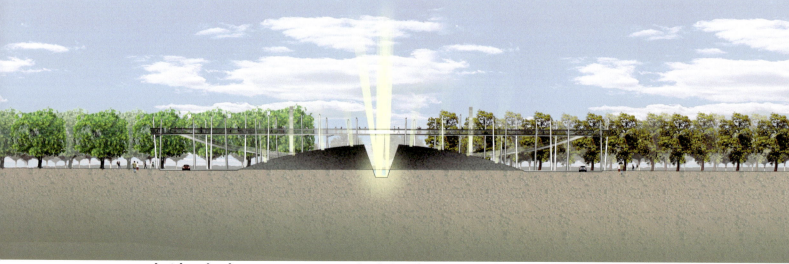

pedestrian sky ring   central axis landform   elevated ring for viewing

Element: **EARTH**
元素：土
Olympic Color: **BLACK**
奥林匹克颜色：黑

奥林匹克广场
Gateway Plaza

54

Element: **FIRE**
元素：火
Olympic Color: **RED**
奥林匹克颜色：红

Plan
平面图

Sketch View of Sport Plaza Gateway Fountain
体育广场的入口喷水展示

体育广场
Sports Plaza

flame fountain
焰火喷水展示

Element: **WATER**
元素：水
Olympic Color: **BLUE**
奥林匹克颜色：蓝

Plan
平面图

Sketch View from Water Theater
水剧场透视图

水剧场新闻广场
Media Plaza

| Focal Landform | Golden Bridge | Water Feature | Waterfall Landfro |
| 焦点坡地 | 金桥 | 水景 | 水景坡地 |

Plan
平面图

Element: 金 METAL
Olympic Color: 金 GOLD

Sketch View from Plaza toward Golden Bridge
从广场望向金桥

Garden Landform and Kiosks
地景设计和商亭

Water Theater
水剧场

Strolling Garden
园景

文化广场
Cultural Plaza

57

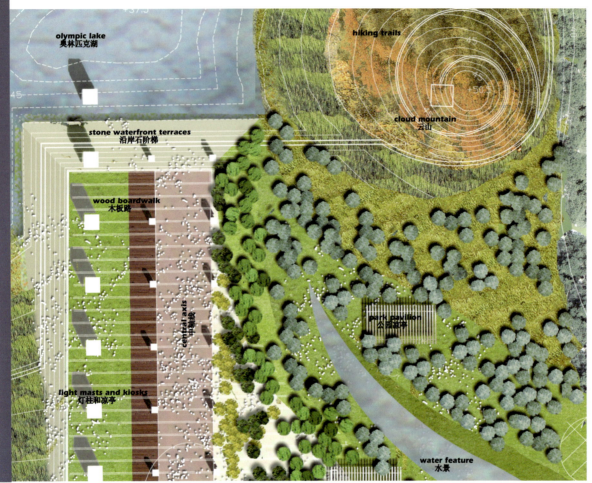

Plan
平面图

Element: WOOD
元素：木
Olympic Color: GREEN
奥林匹克颜色：绿色

森林广场城市中轴线融入大自然
Forest Plaza

Element: **WOOD**
元素：木
Olympic Color: **GREEN**
奥林匹克颜色：绿色

森林广场和城市
节奏相互平衡
Natural Systems in Harmony With Urban Rhythms

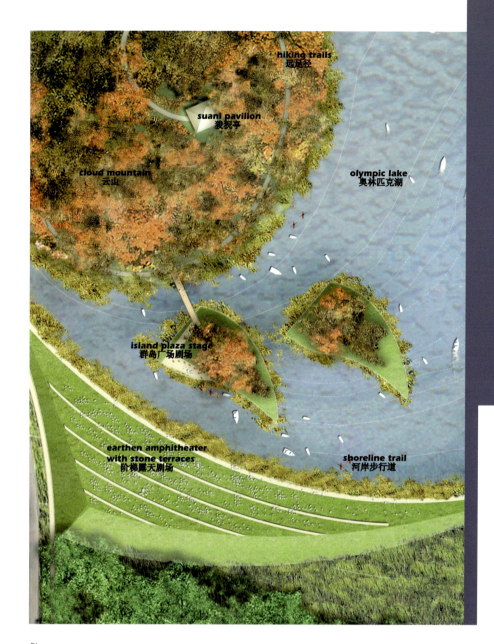

Plan
平面图

焰火，喷水展示，旗帜飘扬

suani pavilion on cloud mountain
在云山上的狻猊亭

olympic lake
奥林匹克湖

stone and grass amphitheater terraces
石和草形成的阶梯露天剧场

59

游览路线图
Tourist Map

ROAD HEIRACHY 道路层级

TRANSIT 公共交通

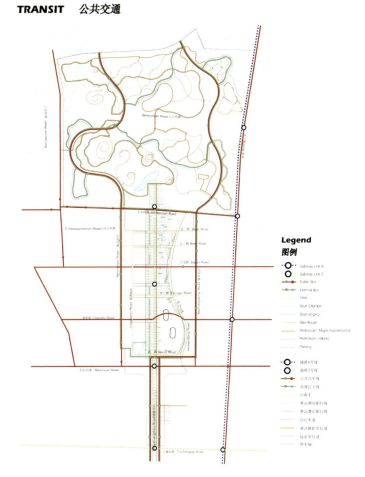

ROAD SECTIONS 道路剖面

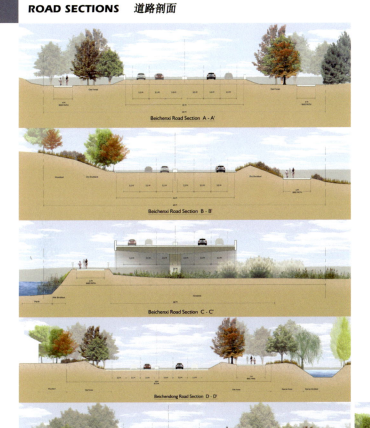

SECTION KEY 剖面位置示意

道路交通图
Circulation Map

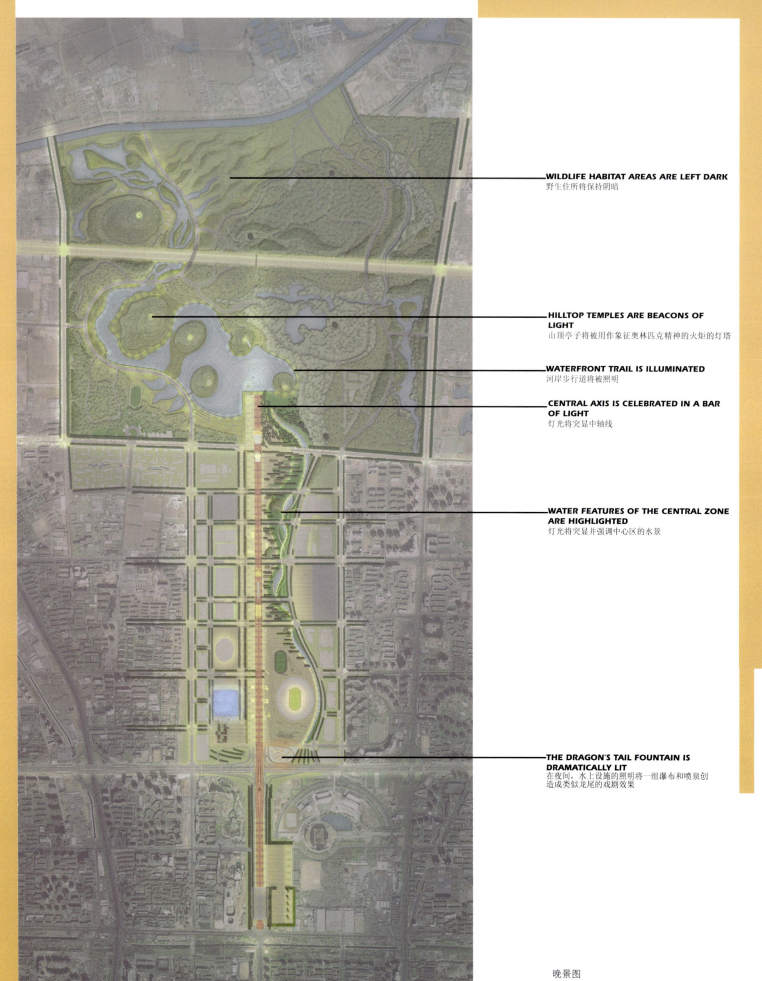

WILDLIFE HABITAT AREAS ARE LEFT DARK
野生住所将保持阴暗

HILLTOP TEMPLES ARE BEACONS OF LIGHT
山顶亭子将被用作象征奥林匹克精神的火炬的灯塔

WATERFRONT TRAIL IS ILLUMINATED
河岸步行道将被照明

CENTRAL AXIS IS CELEBRATED IN A BAR OF LIGHT
灯光将突显中轴线

WATER FEATURES OF THE CENTRAL ZONE ARE HIGHLIGHTED
灯光将突显并强调中心区的水景

THE DRAGON'S TAIL FOUNTAIN IS DRAMATICALLY LIT
在夜间，水上设施的照明将一组瀑布和喷泉创造成类似龙尾的戏剧效果

晚景图
Nightscape Map

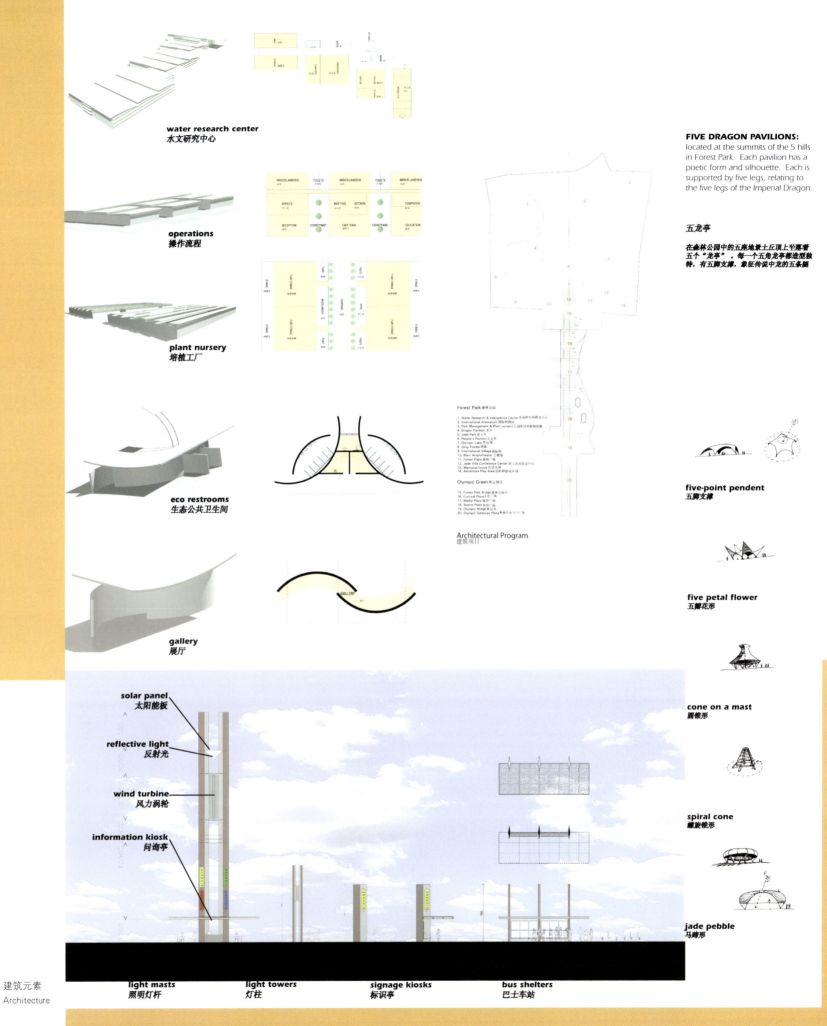

SUSTAINABLE FEATURES
可持续性

Environmental Education
环保教育
- Sustainable Building Showcase 可持续性建筑设计的展示窗口
- International Arboretum 国际植物园
- Children's Garden 儿童花园
- Nature walks 自然步道
- Water Research & Interpretive Center 水文研究教育中心

Alternative Energy and Energy Efficiency
可替代能源以及节能措施
- Solar-powered bus station 太阳能巴士车站
- Solar-powered lighting 太阳能照明设施
- Energy efficient fixtures and equipment 节能装置
- Energy minimizing building design 节能建筑设计
- Wind generated power 风能
- Geothermal 地热

Alternative Transportation
交通方式
- Link to subway 与地铁线相连
- Electric shuttle buses 电力巴士车
- Bike racks 提供自行车存放架
- Extensive trails 广泛丰富的小路

Water Efficiency and Storm Drainage
节水和泄洪
- Low water-use fixtures 节水装置
- Onsite treatment of graywater 实时中水处理
- Reuse of graywater 中水再使用
- Stormwater recharge of rivers and aquifers 河道和地下含水层的补水
- Pervious pavement 透水性良好的铺地

Sustainable Landscaping
可持续性景观设计
- High CO$_2$ absorption 强力吸收二氧化碳
- Low water use 节水的设计
- Low maintenance 维护要求低
- Indigenous/adaptive plants 本土的/适应性强的植被
- Construction monitorings 施工期间的环境监控

Sustainable Materials
可持续性的材料
- High recycled content 多使用回收材料
- Locally available 地方材料
- Rapidly renewable 快速更新
- Low emitting 低反射表面

Sustainable Operations
可持续性的运作
- Recycling 材料回收再利用
- Green cleaning 绿色清洁处理
- Integrated pest management 综合的虫害治理方式
- Energy operating plan 能量使用方案

LANDSCAPE ELEMENTS
景观设计基本元素

Existing Vegetation	现有植被
Reforestation	重新造林
Vegetative Rehabilitation	植被复原
Riparian Landscape	流水植被
Water	水面
Transplanted Tree Zones	林木移栽地带

CULTURAL ELEMENTS
人文因素

Existing Archeological Site	现存历史遗迹
Existing Memorial Grove	现存纪念林
Proposed Cultural Elements	设计中的人文因素
Cultural Landscape	文化景观

北京市已经在改善空气和水质方面作了大量工作，目前正在投入巨资发展可再生能源。2008年奥运会为北京提供了向世界展现其环保努力以及环保设计领域的最新发展的机会。对环境负责的奥林匹克公园是奥委会留给北京的最重要遗产之一—它将让人们永远记住富有智慧的设计、先进的技术以及人的创造力可以在繁忙的城市环境中创造一片生态的乐土。

Beijing as a city has made long inroads in improving its air quality and water quality, and is investing heavily in developing renewable energy sources. The 2008 Olympics will provide the city the opportunity to demonstrate to the world its commitment to environmental stewardship and showcase the latest innovations in environmental design. An environmentally-responsible Olympic Green is one of the most important legacies the Committee can leave for Beijing— it will forever serve as a reminder of how intelligent design, sophisticated technology, and human ingenuity can combine to create a man-made ecological utopia within a busy urban setting.

环境规划图
Environmental Planning

河岸设计图
Shope Scope

Central Zone 中心区

Beichengxi Road 北城西路

Amphitheater 露天剧场

Forest Plaza Water Stair 森林广场沿岸阶梯

Boat Dock 船坞

Interpretive Trails 解说径

Cleansing Water Edge 净水河岸

Pond (Flood Water Cleansing) 暴水净化池

Cleansing Water Edge 净水河岸

Planting Strategy

植物分布图

Winter: Magnolia denudata, Mulan, blossoms in the late winter throughout the Cultural Plaza area
冬：文化广场的白玉兰在晚冬时节盛开

Spring: Prunus yedoensis, Yoshino Flowering Cherry, sends forth light pink blossoms in April in the zone of the Media Plaza
春：新闻广场中的樱花在四月间绽放

Summer: Koelreuteria paniculata, Goldenrain tree, produces yellow flower blossoms in the summer throughout the Sports Plaza area
夏：栾树在盛夏把体育广场染成金黄

Autumn: Catalpa bungei, Beijing Catalpa tree, turns yellow in the fall and loses its leaves. Catalpa leaves were once valued at autumnal equinox ceremonies.
秋：璀璨的楸树是金秋的象征

Axis: Sophora japonica, Chinese Scholar Tree, lines the west side of the axis, and has yellow summer flowers.
中轴线：国槐在中轴线西侧一字排开，在夏季飘散幽香

Suitability: in each garden, each particular thing - plant, pavilion, rock, water body - has only one place where it belongs. The gardener does not think of how to design and add to space, rather than the other way around; the gardener's most important skills are perception and interpretation, not conception.

适宜性：中国传统的古典园林设计，讲究的不是如何将设计加入空间之中，而是讲求如何让自然得到表达，追寻"虽由人作，宛若天成"的境界。计成（1582-1634），中国明朝的造园专家，中国古典园林设计的鼻祖，在其专著《园冶》中，提出了园林设计"巧于因借，精在体宜"的设计原则，即因地制宜，而灵活配置。

Succession: ecological succession is demonstrated by the arrangement of various plant "zones" based on soil moisture, exposure, and terrain. Plant communities range from wetland systems to upland forests.

延续性：植物设计概念充分体现了生态延续的规律，根据不同的土壤湿度、朝向和地形的情况而划分出从沼泽、近水、陆生和丘陵等不同的植物类型区。

Season: street trees are planted to highlight the seasons. Four zones are distinguished by doing so: the Olympic Gateway, the Sports Plaza, the Media Plaza, and the Cultural Plaza. The axis itself is distinguished by a continuous, consistent line of planting.

季节性：道路两侧种植行道树突显四季的变化，这样处理的区域有四个：奥林匹克入口广场、运动广场、新闻广场和人文广场。

Grassland: dry upland area that changes color seasonally
草原：干燥的高地，季节性的变换颜色

Dry Meadow: flowering dry upland area that emphasizes the Dragon's mane
草场：开花的干燥高地植物，平面分布形似"龙发"

Shrubland: transitional plant community adding height and interest to the landforms
灌木区：过渡的植物群落，高度变化增加了地形的趣味性

Woodland: open canopy forest consisting of small stature, early emergent trees
林地：开敞林地，包括小型乔木和早生树种

Oak Forest: closed canopy forest located at higher elevations along the perimeter of the site
橡树林：布置于项目外围较高海拔处的大型乔木

Pine/Oak Forest: dense mixed forest found on northern slopes
松树/橡树林：北侧密集混交林

Spruce/Fir Forest: dense, evergreen forest found at high elevations
杉树林：分布在高处的密集常绿针叶树林

Oak Forest & Shrubland: cultural plantings emphasizing landforms & seasons
橡树林及灌木：突出地形变化与季节变化的文化种植

Riparian Forest: open forest community found adjacent to water bodies
滨水林地：水岸的开敞林木群落

Riparian Shrubland: transitional community occupying stormwater retention basins
滨水灌木丛：位于雨洪过滤池的过渡型植物群落

Shallow Wet Meadow: wetland planting tolerating fluctuating water levels and filtering stormwater
浅湿草地：耐湿的沼泽植物，又用于过滤洪水

Deep Emergent Marsh: continuously inundated wetland planting serving as part of the water treatment train
深滞泽湿地：持续性积水处理水处理净化试段

春 Spring

夏 Summer

秋 Fall

冬 Winter

季节图
Seasons

A02
Sasaki Associates, Inc.

Beijing Tsinghua Planning Corporation
北京清华城市规划设计研究院

奥林匹克公园森林公园及中心区景观规划设计方案

OLYMPIC GREEN LANDSCAPE DESIGN FOREST PARK AND CENTRAL ZONE

总体规划图
Overall Master Plan

SUMMARY

概述

以往的奥林匹克会场的选址往往选择远离城市中心的位置或散置于城内,北京奥林匹克公园与此不同,其选址与城市紧密结合,处于一个战略性位置。北京奥林匹克总体规划旨在满足奥运会场馆功能的基础上,赋予北京城中轴线新的延伸和内涵。我们的景观设计方案延续了总体规划的理念,同时对关键景观要素进行了深化设计。

我们之所以将方案命名为"通向自然的轴线",是因为这个名词体现了我们的设计主旨和理念。

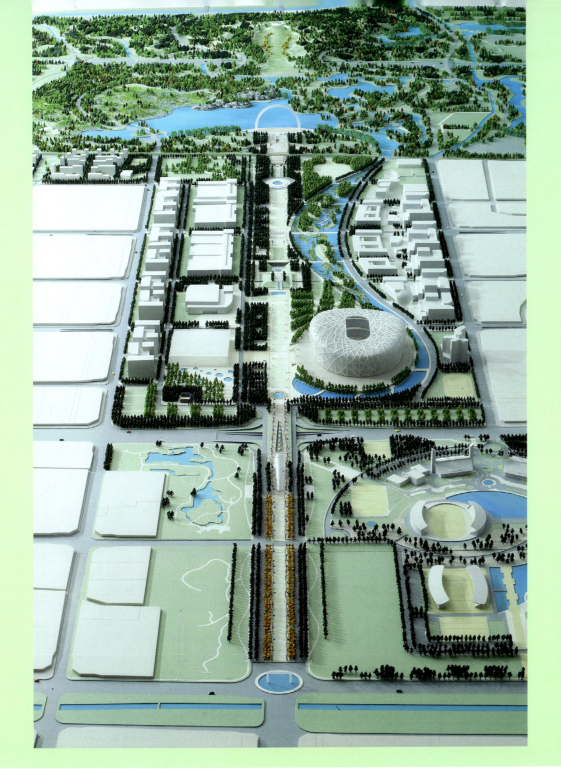

The Beijing Olympic site is strategically positioned and integrated into the existing city, unlike previous Olympic Games in other cities, which were by necessity either scattered throughout multiple areas, or built at distances from the heart of the urban core. The Olympic Green master plan recognizes the unique opportunity to construct a precedent setting Olympic venue - one which can also be a new, powerful and enduring landscape to complement and extend the Beijing Central Axis. Our plan for the site responds to the opportunities of the Olympic Green, and amplifies the elements of the master plan through the crucially important landscape design.

We term our design proposal "Axis to Nature". Even the name is a reflection of the many dualities that guide and are a part of this design.

DESIGN CONCEPT
设计理念
通向自然的轴线

（一）．过去·未来

对历史的研究可为北京未来的景观规划提供依据。在城市规划史上，中国对世界城市的发展有着不可磨灭的贡献，传统的中国城市总体规划法则在当今仍然对世界城市的发展具有重要意义。

中国历史上的大规模新城规划多采用棋盘式布局，体现出对秩序的追求。对秩序与对和谐的追求在中国传统园林设计上也得到了完美的体现。中国传统园林设计力求与自然和谐统一，表达了对自然的尊重。同时，阴阳相生相克的法则也普遍应用于中国园林设计。我们在设计方案中运用了这些中国传统的设计理念，采用大胆的现代设计手法演绎中国传统设计精髓，打造明天奥运的壮美蓝图。通过对北京本土动植物和生态系统多样性的研究，我们制定了方案指导将这一庞大景观系统建设成为有可持续发展能力的、能涵养生物多样性的自然生态系统。

（二）．南·北

北京城内有许多世界建筑和园林史上的佳作，天安门广场与气势磅礴的紫禁城确定了北京城中轴线的南部重心，沿中轴线附近坐落着举世闻名的景山、北海、天坛，每个古建筑群都气魄宏大，体现出中国古代建筑与园林艺术的完美结合。奥林匹克绿色公园的设计在北京城中轴线的北端形成了一个纪念性标志，与南端的标志性古建筑群相互呼应，和谐平衡。

（三）．城市·自然／曲·直

作为北京城中轴线向北的延伸，奥林匹克公园的景观设计延用了北京城的规划与景观设计原则，城市棋盘式网格通过贯穿公园南北轴的有序景观被延伸至公园内。这套网格起自湖畔，止于森林公园广场，在这里"城市融入自然"。这个21世纪的城市广场是举行集会的理想场所，与中轴线南端的历史城市中心天安门广场南北呼应，达到平衡。

曲线形的水面彰显了几何直线的城市格局，同时引入了自然的形态。森林公园之湖形似蟠龙，北起森林公园，向南面的国家奥林匹克主会场延伸。在南端，水面如盘旋的龙尾环绕着国家体育馆，衬托其标志性的地位。国家游泳中心边的圆形喷水池是水系的终点，好似龙尾激起的浪花。

Axis to Nature

1. Past and Future

In designing the future landscape of The Beijing Olympic Green, it is informative to consider the past. In city building, China has provided the world with timeless and enduring principles that remain relevant today.

Guided by the timeless idea of the grid, Chinese city planning embraces the principle of civic order, formed in large part by the creation of open spaces planned and executed at a grand scale. This sense of order and of seeking balance is manifested in the landscape and garden traditions of China. This special respect for the landscape has evolved a design expression with a unity and harmony with nature. The idea of paired opposites is an integral part of Chinese landscape design. Each of these Chinese principles is employed in our design, interpreted in a bold, contemporary way, using respect for the classical designs of history to move towards the future. Similarly we use knowledge of the past (former) ecological diversity of Beijing's plants and animals to build a more sustainable, richer environmental future for this landscape.

2. South and North

Beijing is the location of some of the world's most significant examples of city building and landscape design. Tianemen Square and the great courtyards of the Forbidden City anchor the southern end of Beijing's central axis. Nearby, The Summer Palace, Bei Hai Park, and the Temple of Heaven are internationally recognized landmarks. Each is characterized not only by the integration and harmony of architecture and the landscape, but also an appropriate epic scale. Our design for Olympic Green creates a contemporary monument on the northern end of the central axis, to balance and harmonize with these great monuments at the south end of the axis.

3. City and Nature/ Straight and Curving

As the northern extension of the city's axis, our proposal for the landscape design of Olympic Green has continuity with these fundamental principles of city planning and landscape design. The city grid is extended into Olympic Green area through an ordered landscape that extends the north-south axis. This grid extends to the edge of the lake and arrives at Forest Park Square, where the city meets the created natural environment of Forest Park. This large civic Square is a great gathering space created in the 21st century as the northern counterpart to the historic squares and courts in the center of the city.

The geometry of the city is amplified and juxtaposed with the curving shapes of water, reflecting the natural patterns. Forest Park Lake, the curving dragon-shaped body of water, extends southward from Forest Park towards the principal venues of Olympic Green. The water concludes at the National Stadium, where the southern end of the body of water (the tail of dragon) wraps around the National Stadium to embrace the landmark structure. The round fountain pools adjacent to the National Swimming Center are positioned at the end of the water course (as the turbulence from the tail of the dragon).

(四). 天·地

北京城轴线上的城门序列和朝向有其相应的含义与用途，服务于特定的仪式与活动。奥林匹克公园主轴上的大门运用了相似的原则，轴线南端为"奥运之门"，是一座高40m的不锈钢和铝的方拱形体，跨越于位于中轴线与元大都城墙相接的圆形的倒映池上，水中方拱的倒影与方拱本身形成一闭合的正方形，这座"地方"之门位于奥林匹克中轴线最靠近城市的端点。

奥林匹克公园5km长的中轴线成为标记中华五千年文明史自然环境艺术成就的理想场所。依照从中华文明的起源到历朝历代，至新中国成立以来的50年，直至无穷未来的发展顺序，中轴线被分成5段1000m长的组成部分，象征着5个主要历史时期。每段轴线上设以其代表的朝代为主题的纪念性建筑、广场、大型景观或大地艺术。在轴线末端，象征着未来的森林公园由此展开，我们将运用生态学原理恢复北京自然环境与生物多样性，用以表达北京人民对完美和富有活力的未来的追求，也标志着北京地区大规模生态环境改造的开始。

轴线向北延伸3.5km处为森林公园广场，入口标志为一凌越于方形基座之上的半圆拱，它的灵感源于传统的中式月亮门。99m高的金属半圆形森林公园之门将城市中轴线与森林公园中轴线连接起来。与"奥运之门"相同，龙形湖面中"森林公园之门"的倒影与其本身形成了一个完整的圆环，是沿城市中轴线的景线与沿森林公园水系横向轴线景线的交叉点，表达了"天圆地方"的古老又现代的人文设计理念。

(五). 山·水／纵·横

透过半圆形"森林公园之门"北望，是中国传统自然山水的湖山写意。山水塑造上采用挖湖堆山传统手法，开掘80hm²的水系，多余土方用于北界屏障的山体叠造，龙形湖体的北界山形陡峭，形成了与水平向的开阔水面的鲜明对比。连绵的缓坡平滑地过渡到北边更高的圆形小丘处。在接近地势最高点为一椭圆形的纪念广场，标志着新中国的50年，预示着未来的发展。

(六). 艺术·科学／人·自然

奥林匹克绿色公园为艺术与科学的结晶。我们的设计既运用抽象的几何概念，取材于中国传统文化，同时与中国环境艺术的设计原则有密切的联系。森林公园是人为的生态公园，旨在增加城市生态的多样性。森林公园的宏大规模为创建不同类型的景观区提供了条件，从三种不同形态的湿地、缓坡到高地森林景观。这些生态环境有利于恢复本土动植物群落，使城市的居民有机会亲近自然。公园里有供游客使用的场所，也设有生态保护禁区，例如湖心岛上设鸟类栖息保护区。野生动物保护区将带动其周边地带发展成为更有活力的生态环境。

虽然设计沿用了中国传统理念，但同时运用了现代的设计手法以适应21世纪新北京的形象要求。这种现代的演绎手法体现了创办高科技奥运、尊重自然、创北京新形象的多重目标。与以往的奥运会会场不同，北京奥运人中心与城市有机地交织在一起，不仅在奥运会举办期间成为城市的焦点，而且在奥运会之后还会继续延续其在城市中的功能。

我们的设计过程可形象地比喻为创作中国国画。在挥笔作画之前，画家对其对象进行仔细观察，做到胸有成竹；我们的设计起源于体现作品神韵的抽象理念。画家在作画时，运笔自如，一挥而就，观者透过其鲜活的笔触体会到画家作画时心中的意境；在我们的设计中，"通向自然的轴线"连接城市与自然，沿"通向自然的轴线"的景观要素表达了设计的抽象理念，如门、广场、中国传统园林的永恒主题——水。

景观的设计的表现方法可被比作为水墨写意。我们着重于表达主题概念的意境，正如简约灵活的写意山水，并从这个方案构思中演绎出细节设计。

Nature

4. Earth and Heaven

Along the existing Beijing axis, the idea of the city gate or entry facilitates orientation and expresses the movement between distinct spheres of activity. So too with Olympic Green. Centered on the main axis, the Southern entrance is The Olympic Gate, a large square steel and aluminum structure spanning forty meters. Only half of The Olympic Gate is built. The other half of the square is created by its reflection in a large oval pool centered on the axis adjacent to the old city wall. This gate represents the earth, closest to the center of the city.

Along the 5 kilometer axis through the Olympic Green we commemorate the achievements and contributions of 5,000 years of Chinese environmental design. Beginning with the early development of settlements, through the great dynasties and the last 50 years, and moving into the future, the axis is divided into 1,000 meter segments symbolizing 5 major periods in Chinese environmental history. Each period is commemorated in a site designed as a plaza, landform, or larger landscape setting. The final period, representing the future, opens into Forest Park, where ecological principles are used to restore the rich biodiversity of Beijing's natural environment. This being a living expression of a more complete, dynamic future for the people of Beijing, and the beginning of a great environmental improvement in the surrounding region.

Three and half kilometers north at Forest Park Plaza, the entrance to Forest Park is marked by The Circle above the Square. Inspired by the idea of a moon gate, the Circle is a 99 meter high metal half-circle gate from the axis of the city to Forest Park. Like the Olympic Gate, only one half of The Circle is built. The other half is the reflection in the dragon shaped lake. The view through The Circle connects the city and the north-south axis to Forest Park.

5. Water and Hills/ Horizontal and Vertical

The view through the Circle looks north into Forest Park, an idealized Chinese landscape of hills over water. Fill from the excavation of the 80 hectare dragon shaped lake will be used to create hills 60 to 90 meters high. The southernmost hill is designed with steep topography, to create a dramatic juxtaposition with the dark horizontality of the lake. Beyond, a gently sloping meadow extends from the water to taller rounded hills. Near the highest point, an oval commemorative plaza marks the last fifty years and points to the future.

6. Art and Science/ Man and Environment

The Olympic Green landscape is a product of both art and science. The design reflects abstract geometric ideas, and includes metaphorical references to Chinese culture. But equally as important, our proposal is deeply connected to the principles of environmental design that have evolved in China. Forest Park is a created ecological park to increase the biodiversity of the city. The size of Forest Park creates an opportunity for multiple zones, ranging from three types of wetlands, meadows, to upland forests. These environments will allow reintroduction of indigenous plants and animals within reach of Beijing's people. The park will not only have areas for people to use, but also places with restricted access primarily focused on ecology, like rookeries on islands to support bird habitats. These zones for wildlife will become the centers from which the wider region becomes more sustainable with dynamic, living elements of nature.

While the design has continuity with the past, these ideas are interpreted in a distinctly contemporary approach, in a way that is appropriate for Beijing in the 21st Century. This modern interpretation reflects the dual aims of making a high tech Olympics with an environmental consciousness, and a new city image for Beijing. The Olympic Green is integrated into the city fabric unlike any other Olympic site, creating a powerful city image during the games. But the Olympic Green, like other great landmarks, is designed to endure for centuries.

Our design has been developed in a process like traditional Chinese painting. In this tradition, before starting a work, the artist observes the site very closely, and develops an overall concept (the conception). This concept is developed with abstract ideas that describe the spirit of the work. The painting is then made as a freehand expression of the concept with the observer able to imagine the original appearance and realize the meaning. The concept of the "Axis to Nature" links the dualities of city and nature at Olympic Green. The landscape design contains a series of abstract ideas along the "Axis to Nature", such as gates, squares, and water derived from the timeless traditions of Chinese design. The design illustrated in our work is like a freehand expression of the concept. It shows the main ideas, but like the brush work in a painting, the emphasis in this proposal is on the concept and its meaning. The supporting details will follow logically from this plan.

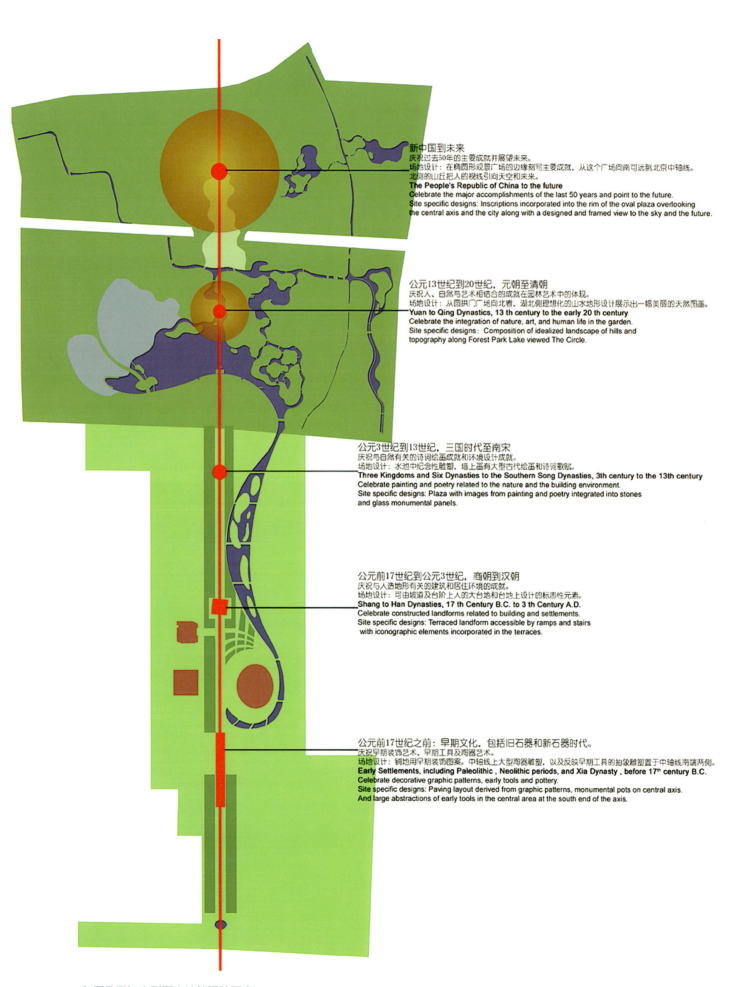

新中国到未来
庆祝过去50年的主要成就并展望未来。
场地设计：在椭圆形观景广场的边缘刻写主要成就。从这个广场向南可远眺北京中轴线。
北侧的山丘把人的视线引向天空和未来。
The People's Republic of China to the future
Celebrate the major accomplishments of the last 50 years and point to the future.
Site specific designs: Inscriptions incorporated into the rim of the oval plaza overlooking
the central axis and the city along with a designed and framed view to the sky and the future.

公元13世纪到20世纪，元朝至清朝
庆祝人、自然与艺术相结合的成就在园林艺术中的体现。
场地设计：从圆拱门广场向北看，湖北侧理想化的山水地形设计展示出一幅美丽的天然图画。
Yuan to Qing Dynasties, 13th century to the early 20th century
Celebrate the integration of nature, art, and human life in the garden.
Site specific designs: Composition of idealized landscape of hills and
topography along Forest Park Lake viewed The Circle.

公元3世纪到13世纪，三国时代至南宋
庆祝与自然有关的诗词绘画成就和环境设计成就。
场地设计：水池中纪念性雕塑，墙上画有大型古代绘画和诗词歌赋。
Three Kingdoms and Six Dynasties to the Southern Song Dynasties, 3th century to the 13th century
Celebrate painting and poetry related to the nature and the building environment.
Site specific designs: Plaza with images from painting and poetry integrated into stones
and glass monumental panels.

公元前17世纪到公元3世纪，商朝到汉朝
庆祝与人造地形有关的建筑和居住环境的成就。
场地设计：可由坡道及台阶上人的大台地和台地上的标志性元素。
Shang to Han Dynasties, 17th Century B.C. to 3th Century A.D.
Celebrate constructed landforms related to building and settlements.
Site specific designs: Terraced landform accessible by ramps and stairs
with iconographic elements incorporated in the terraces.

公元前17世纪之前：早期文化，包括旧石器和新石器时代。
庆祝早期装饰艺术、早期工具及陶器艺术。
场地设计：铺地用早期装饰图案。中轴线上大型陶器雕塑，以及反映早期工具的抽象雕塑置于中轴线南端两侧。
Early Settlements, including Paleolithic, Neolithic periods, and Xia Dynasty, before 17th century B.C.
Celebrate decorative graphic patterns, early tools and pottery.
Site specific designs: Paving layout derived from graphic patterns, monumental pots on central axis.
And large abstractions of early tools in the central area at the south end of the axis.

奥运公园与中心区中轴线设计示意 Axis Diagram of Olympic Green and Central Zone

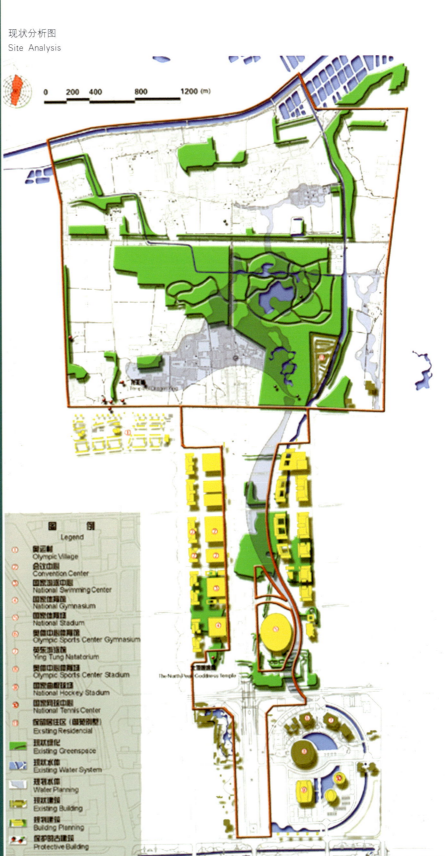

现状分析图
Site Analysis

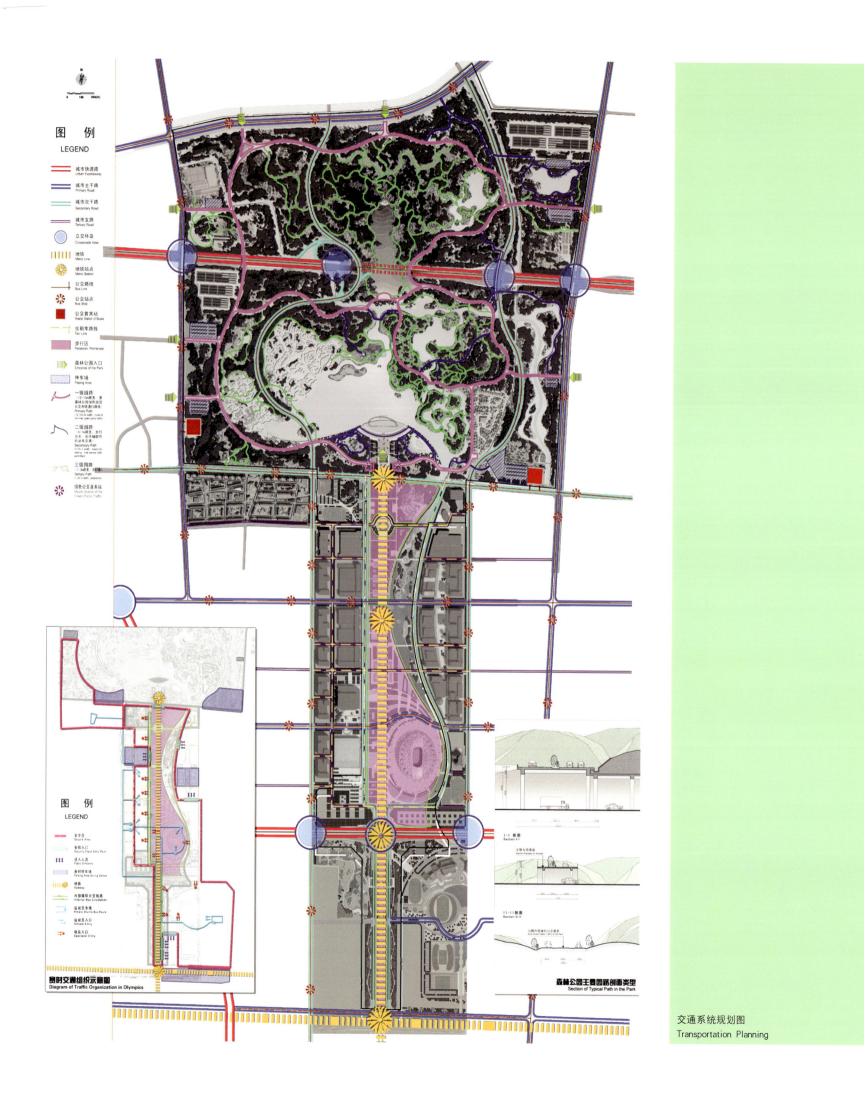

交通系统规划图
Transportation Planning

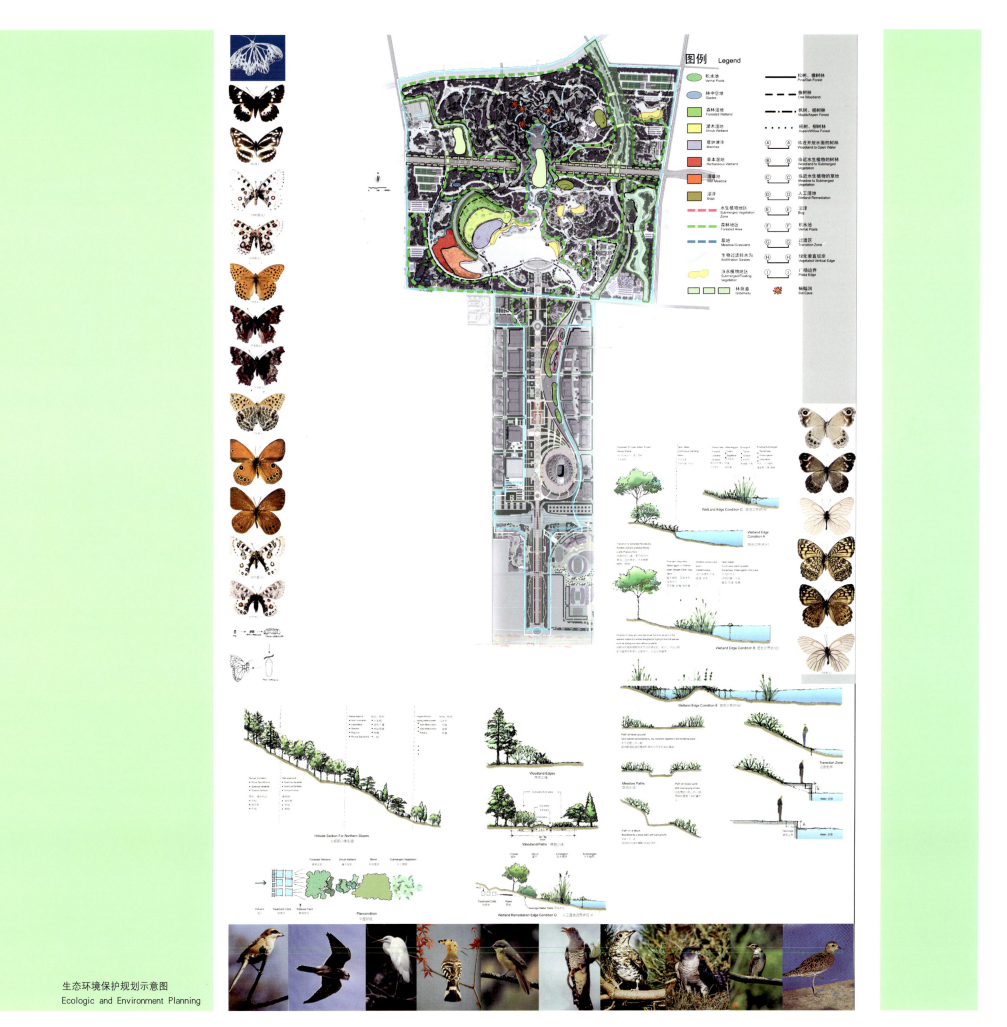

生态环境保护规划示意图
Ecologic and Environment Planning

Axis to Nature

森林公园总体规划图
Forest Park Overall Master Plan

Axis to Nature

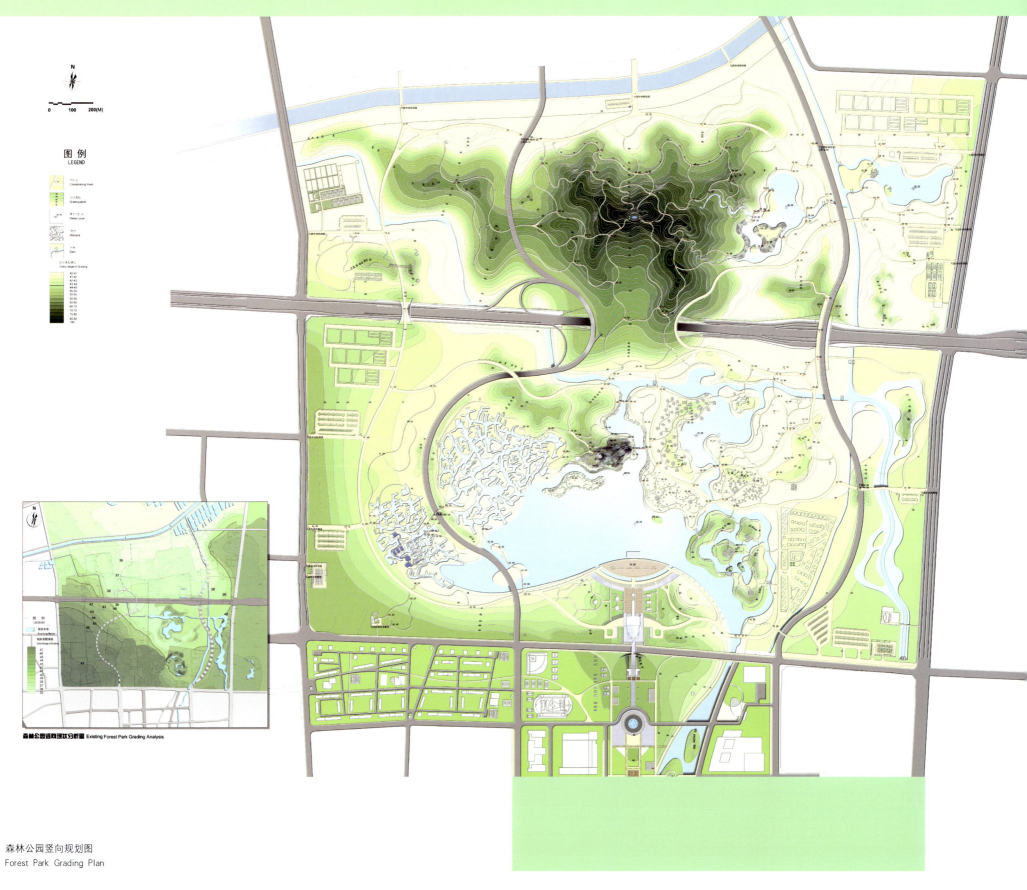

森林公园竖向规划图
Forest Park Grading Plan

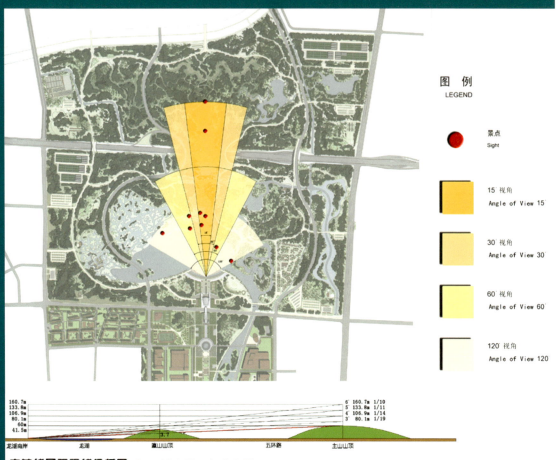

中轴线景观视线分析图 Central Axis View Analysis Plan

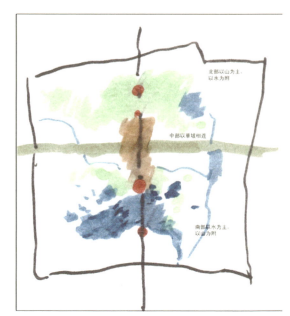

Mountain and Water Structure Design Concept 山水间架结构设计构思

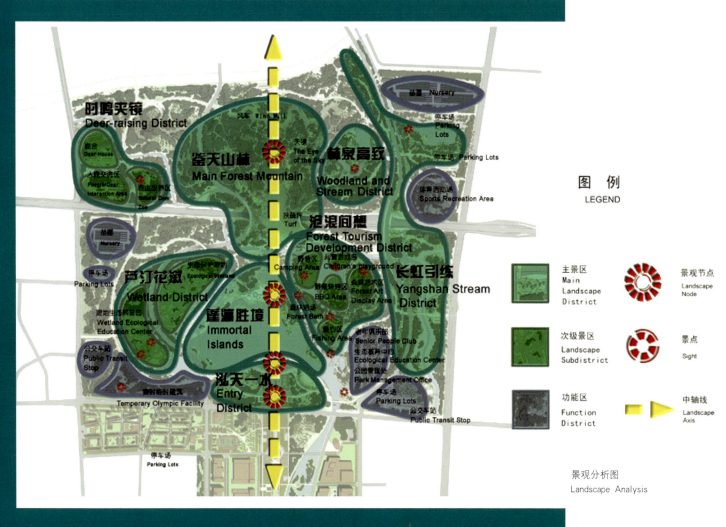

景观分析图
Landscape Analysis

湿地科普园 1-1 剖面 Wetland Park Section 1-1

湿地科普园 2-2 剖面 Wetland Park Section 2-2

湿地科普园平面图 Wetland Science & Education Park Plan

湿地公园平面 Wetland Park Plan

野营区 1-1 剖面 Camping Area Section

湿地科普园 3-3 剖面 Wetland Park Section 3-3

湿地科普园 4-4 剖面 Wetland Park Section 4-4

本规划通过模拟地表径流对高程变化较缓的地面冲刷后的地貌进行湿地设计。

根据原有地形由西南向东北倾斜的特点，在北部结合微地形设计，扩大湿地汇水面。同时，湿地与龙形水体紧密联系。从山体到龙湖，蔓延着以放射状为主，纵横交错的水网，水系由疏至密，最后与龙湖相接。

Though in history there is much natural matsh in Beijing especially in north or north-west region now it is not existed.There are a few water systems in Olympic forest Park location,and many studies towards marsh indicate that the key point of artificial marsh is to design formation of marsh So this layout design marsh through physiognomy formed by simulated earth's surface flow washing out high,slower-changed surface.

沧浪间想景区平面 Forest Tourism Development District Plan

　　此景区基址为一个公园，因此现状拥有较好的植被和水体。在设计的过程中希望能充分的利用现状，保留尽可能多的树木，并结合现状布置适当的休闲游憩和文化娱乐设施。如：垂钓区、野营区、野餐烧烤区、森林浴场、儿童游戏岛、森林艺术区（户外雕塑园）等，并设有相应的各类服务设施。

　　This scenic region is a park, so there are many vegetation and water system. In designing process it is hopeful to fully use the status in quo and reserve trees as many as possible and dispose proper laying fallon and cultural amusement, such as whiff area、camping area、picnic barbecue area、forest bathhouse、children playing island, forest art area (outside soulpture park) etc and proper service establishment at all types.

垂钓区 2-2 剖面 Fishing Area Section

天镜景点平面图 Sky Eye Plan

鉴天山林景区总平面图 Mountain Forest Plan

天镜剖面 Sky Eye Section

天镜南立面 Sky Eye South Elevation

天镜北立面 Sky Eye North Elevation

　　作为轴线的端点，奥运森林公园北部以山林景观为主，力求创造出富有自然情趣的生态环境。因此本区避免设置大型的醒目建筑、主要景点有天镜、扶疏坪、风车等。

　　天镜是本景区重要的节点，也是森林公园的制高点。为了维持中轴线终点的自然特征，从湖面、草坪上看它应是"不可见"的，它也应该是一个良好的景观平台。平台中央建有一镜面水池，用于反射天光浮云。这里提供了一处可以感受天与人交流的诗意空间。

As the end of axes, the north of Olympic forest park is mainly designed as mountain and forests in order to create ecological environment with natural sentiment. So this area is avoidable to set large-scale and striking structure. The eye of the sky is this scenic area's important node and also a top point of Forest Park. It`s position determines the speciality in design. In order to keep the natural character of the central axes, it should be invisible from the surface of lake and grass ground, and it should also be a favorable view ground. A mirrored surface pool was built in the center of the ground using to reflect the sight and cloud.

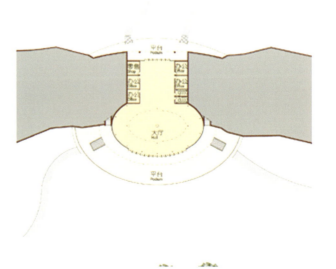

瀛台剖面 1-1　Rock Hill Section 1-1

瀛台剖面 2-2　Rock Hill Section 2-2

　　位于森林公园南部居中的位置，是南区的核心景观。规划力图营造出一个纯粹自然式的完美的视觉景观效果。为入口区的大型拱门创造出自然的优美如画的背景，使北京中轴线渐隐在自然山水中。由于主山距离主入口及湖面太远，已无法起到在视线上强烈的统摄作用，因此在湖北侧规划了一系列配山与岛屿。对于450m的湖面进深和近千米的面阔，我们设计了 41.5m 的山体高度。

This scenic spot is located in the southern center of the Forest Park, being the core landscape of the south area. The planning strives to build a perfect, pure natural visual landscape effect, create a natural picturesque background for the large arch in the entrance region, so that the central axis line of Beijing fades-out in the natural landscape. Since the main mountain is two far away from the main entrance and lake. As for the lake surface 450 meter in width and some 1000 meter in length, we have designed 41.5 meter of forest mountain height.

瀛台意象　Rock Hill Image

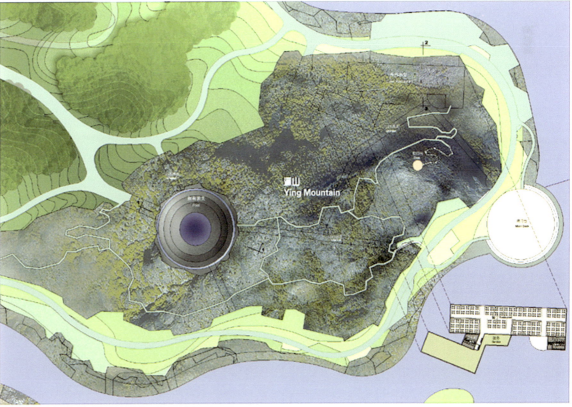
瀛山总平面　Ying Mountain Master Plan

位于五环路以北，主山体东南部，环境相对封闭，处于山体的环抱中。利用在山体向阳的东南部围合形成各种宜人的游览活动空间。山体自然形成谷地设计为两条小溪，从西向东汇入人工湖中，构成森林公园北部山水相依的空间格局。围绕小溪结合小桥，亭台设计一系列自然景观，林阴小径在溪流上穿行形成空间趣味点。

The scene is located north 5th Ring Road, with east south of its main mountain body relative closed and encircled by the mountain. The east south with mountain body facing south, may be enclosed into kinds of delightful activity spaces for sight-seeing. The valley naturally formed from the mountain body is designed into two brooks, converging in an artificial lake, which forms the spacial pattern with mountain and water leaning.

林泉高致景区总平面　Woodland and Srteam District Plan

林泉高致　Beatiful Stream in the woods

林泉高致节点平面 Woodland and Stream District Plan

观景台南立面 Deck South Facade

观景台剖面 Deck Section

观景台平面 Deck Plan

服务中心剖面 Service Center Section

服务中心南立面 Service Center South Facade

服务中心北立面 Service Center North Facade

服务中心东立面 Service Center East Facade

服务中心西立面 Service Center West Facade

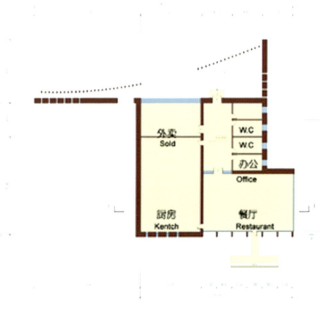

服务中心平面 Service Center Plan

91

"泓水一天"剖面图
Overlook Section

"泓水一天"立面图
Overlook Facade

"泓水一天"平面图
Lakeside Overlook Plan

森林公园水体驳岸线设计图
Forest Park Water Edge Design

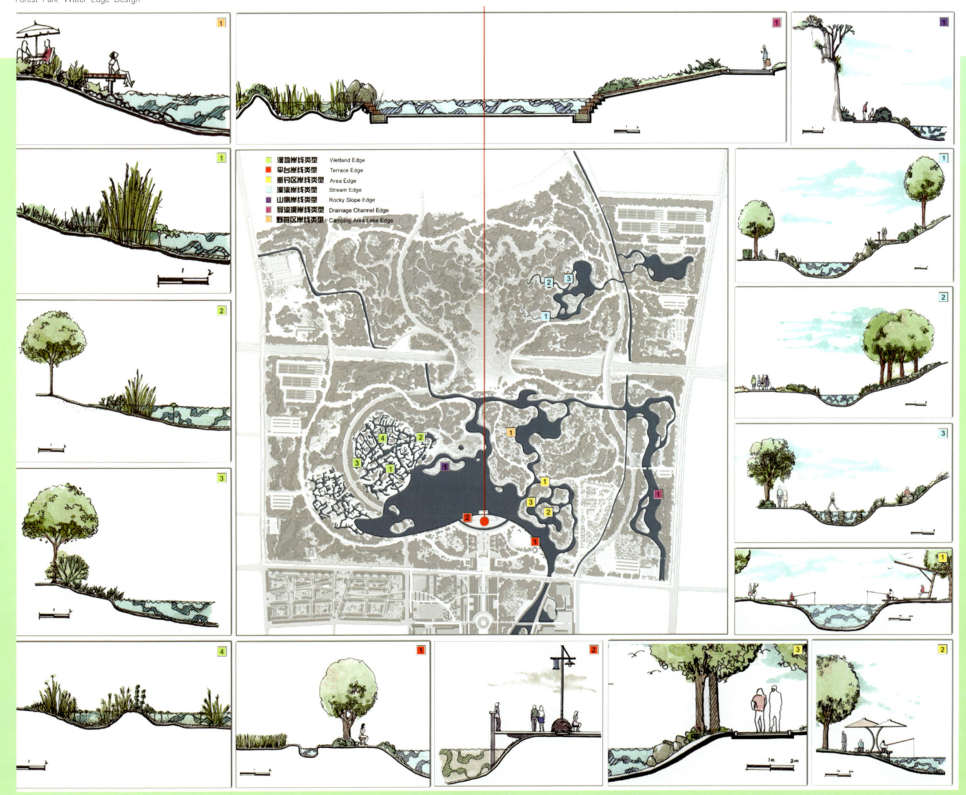

中心开敞空间

主轴为一50m宽，贯穿南北的开阔空间。这条中心开阔地可为各种庆祝表演活动提供场地，在奥运会或其他活动举办期间。中心开阔地可用来搭建帐篷，同时容纳大量人群。延这条主轴每隔1km设一纪念性主题景观，为行人提供步行的节奏感与休憩场所。与紫禁城延中轴线的序列相似，延主轴的视景先以一主题景观为视觉焦点，然后继续延伸。

Central Open Space

There is a 50 meter wide central space, between the trees, from south to north. This central open space is flexible, to allow for the staging of events, performances, and festivals. During the Olympics and other major events, this space could accommodate large tent structures, as well as crowds of people. Within this space, at an interval of one kilometer, a series of focal points are centered along the axis, to create a rhythm of features and resting places. Like the sequence of moving through the Forbidden City, views are centered on each focal point, before moving on.

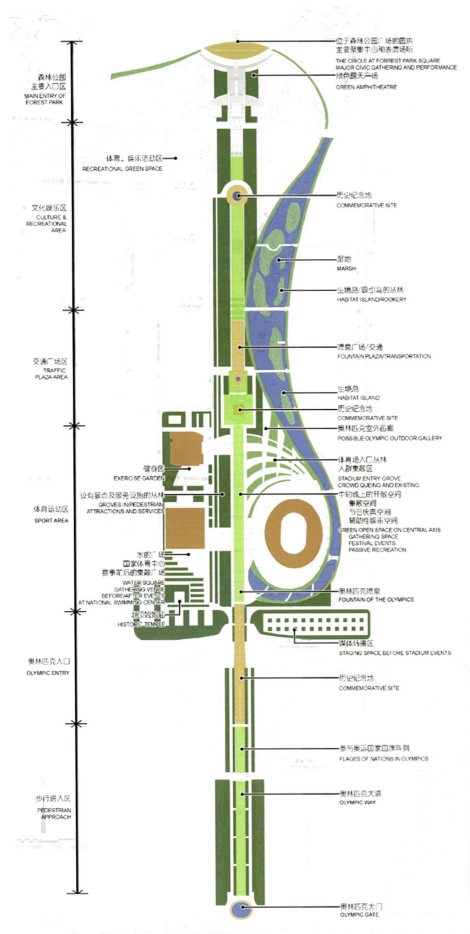

中心区景观规划分析图 Landscape Planning Analysis of Central Zone

Perspective Rendering of Olympic Canal 奥林匹克运河透视图

Axis to Nature

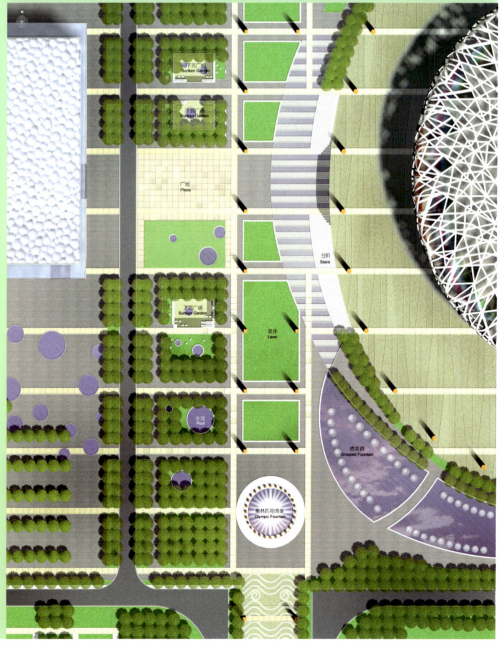

Central Zone

规则式树林透视图 Perspective View in Formal Tree Grove

高台透视图 Perspective View of Raised Terrace

奥运之门透视图 Perspective View of Olympic Gate

奥运之门平面图 Plan of Olympic Gate 1:500

中心区景观节点平面图 Site Plan of Central Area Scenic Spot

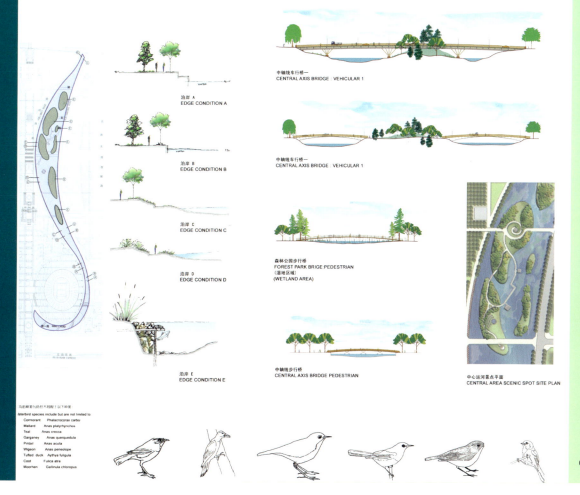

在这一地带规划的植物将最大限度地支持野生动物资源和生态功能，结果特殊植物将在四季吸引各种鸟类，审慎设计的野花将被多彩的蝴蝶所覆盖，中心区将充满自然的音乐而倍显活力，森林公园中的生物将扩散到中心区，然后将扩散到周边的城区。

Plantings in this area will support the theme of maximum resources for wildlife and ecological function. Special plantings of food-bearing trees and shrubs will attract birds in all seasons. Carefully designed wildflower plantings will be covered with colorful butterfies. The axis will be alive with nature's sounds and sites around the beautiful new structures. Life in the Forest Park will spread through the central axis and then out into the surrounding parts of the city.

中心水体岸线设计图
Central Zone Water Edge Design

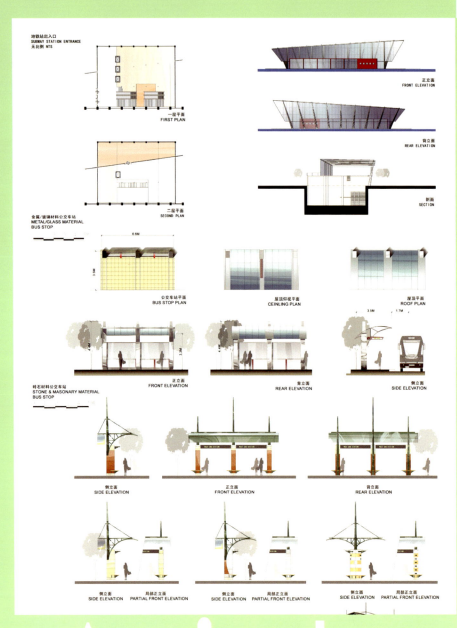

Axis to Nature

A03

北京市城市规划设计研究院
Beijing Municipal Institute of City Planning & Design

Olin Partnership Ltd.
美国欧林景观建筑及城市设计股份有限公司

奥林匹克公园森林公园及中心区景观规划设计方案

OLYMPIC GREEN LANDSCAPE DESIGN FOREST PARK AND CENTRAL ZONE

Dance of Dragon

总平面图
Overall Master Plan

1 元大都遗址公园   Yuan Da Du Historic Park
2 北土城沟   North Tu Cheng Canal
3 奥林匹克公园入口-熊猫环岛   The Olympic Green Entrance-Panda Circle
4 地铁熊猫环岛站   Subway Station(Panda Circle)
5 中华民族园   Chinese Ethnic Culture Park
6 奥体中心   National Olympic Sports Center
7 地铁奥体中心站   Subway Station(National Olympic Sports Center)
8 公共休闲区   Public Recreation Zone
9 北顶娘娘庙   Bei Ding Niang Niang Temple
10 国家游泳中心   National Swimming Center
11 主体育场   National Stadium
12 国家体育馆   National Gymnasium
13 奥林匹克广场   Olympic Square
14 地铁奥运广场站   Subway Station(Olympic Square)
15 会议中心   Conference Center
16 红色飘带   Red Ribbon
17 绿色客厅   Green Living Room
18 花丘   Flower Meadow
19 科技生态示范区   Eco Demonstration Zone
20 汽车电影院   Motor Cinema
21 森林公园入口   The Forest Park Entrance
22 地铁奥运森林公园站   Subway Station(Forest Park)
23 奥运湖   Olympic Lake
24 公共服务区   Public Service Zone
25 公交首末站   Bus Stop
26 园艺博览区   Garden Exhibition
27 清河导流渠   Qing He Canal
28 后勤服务区   Service Zone
29 桃花岛   Peach blossom Islet
30 社会停车场   Public Parking
31 变电体   Transformer Station
32 龙珠山   Dragon Pearl Hill
33 龙珠   Dragon Pearl
34 芳香岛   Aroma Islet
35 大地上的"中国印"   Chinese Logo On The Earth
36 湿地   Wetland
37 青少年生态教育基地   The Youth Eco-educational Center
38 前山   Front Hill
39 后山   Back Hill
40 "瓠"塔   Gu Tower
41 清河   Qing River
42 仰山大沟   Yang Shang Du Canal
43 奥林匹克大道   Olympic Avenue
44 奥运英雄路   Olympic Hero Avenue

DESIGN CONCEPT
设计理念

"赤橙黄绿青蓝紫，谁持彩练当空舞。"
世纪之初，巨龙腾飞。
北京在欢舞。
中国在欢舞。
我们从历史中走来，
我们将走向未来。
在历史与未来之间，我们迎来2008
世界体育的盛会，
人类大同的庆典。

用什么来诉说我们激动的心情？
用什么来搭建欢乐庆典的舞台？
用什么来象征彼岸的精神家园？

是跳动着的"中国印"？
是神秘纯净的"鸟巢"、"水立方"？
是映衬在蓝天、碧水、清风、森林、远山之下的"中轴线"？

还是融汇了这些美丽景观的庆典舞台
——舞动的山水，
——舞动的绿林，
——舞动的笑脸，
——舞动的街市，
——舞动的北京。

舞——最古老的运动形式。
舞——最欢快的肢体语言。
舞——最愉悦的景观梦想。

龙之舞——动感、能量、欢庆——奥林匹克精神的体现。

"Who's there, in a world of colors, dancing with colorful ribbons."
Rising at the beginning of the new century, there it is, a Great Dragon.
Beijing is dancing joyously.
China is dancing joyously.
Through the history have we walked
Toward the future we are marching
Here are we, between the history and the future, embracing 2008,
A Great Event of World Sport
A Great Harmony of Human Celebrations

Through what shall we express our excitement?
Upon what shall we build our arena for this joyous moment?
And to what shall we compare our paradise of spirit?
Should it be the jumping "Chinese Seal"?
the mysterious & pure "bird nests" or "cubage of water"?
the "Central Axis" under the blue sky, by the clean water,
in the fresh breeze and the prosperous forest and
at the view of the distant mountains?

Or should it be this arena of celebrations,
an integration of all the beautiful scenery?

-- The dancing mountains and waters
-- the dancing forests
-- the dancing smiling faces
-- the dancing downtown streets
-- and the dancing Beijing

Dancing --- one of the most immemorial forms of sport
Dancing --- one of the most bright body languages
Dancing --- one of the most joyful sights in our dream

The Dance of Dragon --- a combination of innervation, energy and festivity --- a reflection of the spirit of Olympics.

Dance of Dragon

山与水关系人类生存。中国人自古尊重山水。古人"祭山"而不"损山":"疏水"而不"阻水",可见一斑。我们建城如此,造园也如此,北京如此,中国也如此。

模山范水讲求山水相依,师法自然,因此有"水以石为面","水得山而媚"之说。

Mountain and water is vital to human survival. The Chinese have valued the mountain and water from time immemorial. The ancients "offer sacrifice to a mountain" instead of " damaging a mountain"," dredge water" instead of "blocking water",through which we can conjure up the whole thinking. We build up a city as well as a garden in this same manner in Beijing and in China.

We imitate mountain and water in natural union. Thus People say," Water takes stone as its face and becomes more graceful with the mountain.

The dancing Landscape

山 水 之 舞

Dance of Dragon

　　通过视线分析，如果山的主峰置五环路以北，从南向北眺望视角过低，比例失调，故依据西南高、东北低的地形，将山之高点位于规划湖面的西北，向北渐渐走低，高约20～30m，既增加了山的层次，又显出山的深远，还可以降低土方量。

　　By visual line analysis, placing the main peak north of the fifth-ring road will make the visual angle from south to north too low and out of proportion. The high spot of the mountain is placed northwest of the planned lake surface pursuant to a landform of southwest high and northeast low, lowering little by little northward with the height being about 20~30 meters, which not only adds to the strata sequence and reflects the vastness of the mountain but also reduces the earthwork.

　　以原有规划及现状为依据，适当改动了部分水面，将森林公园的规划奥运湖、湿地与清河导流渠、清河、仰山大沟及原洼里公园内的湖面系统相连，形成了与山林环绕，完整、贯通的水系。湿地池塘的形态源于中国传统的艺术——云纹和农业文化——梯田，蕴含着抽象于自然的形态与功能。

Dance of Dragon

现状地形分析图
Current Landform Assay Chart

规划地形分析图
Landform Planning Assay Chart

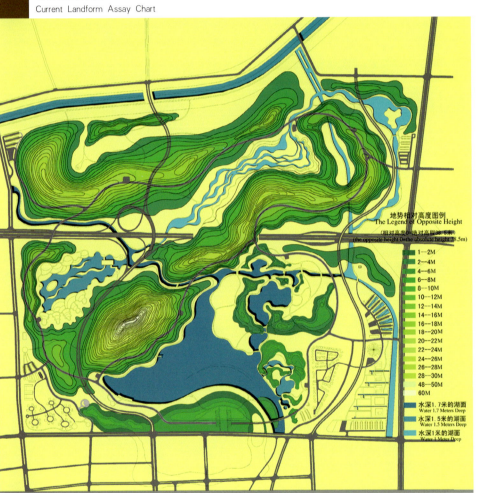

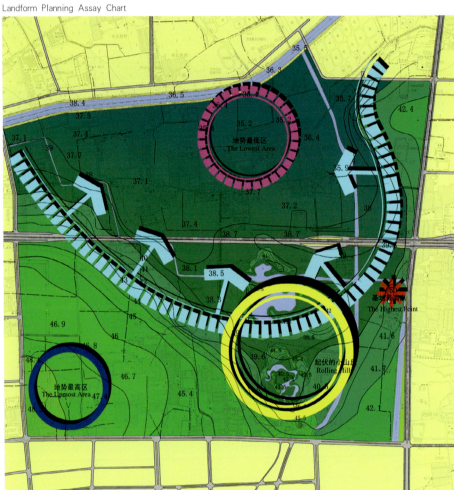

规划遵循原有的地形起伏进行造山挖湖，形成山环湖、溪绕山的景致。所造水形按"龙形"梳理，并使之形成系统，经瘦身的龙脊上置片片龙鳞，将成为北京最活跃、最宜人的滨水场所；所造山形连绵起伏，形成"龙形"，但山峰并不在中轴的延长线上，使北京北部军都山脉能够自然地融入森林公园、融入中轴、融入城市。

The plan follows the original hypsography in making the mountain and digging the lake, resulting in a scenery of mountain embracing the lake and stream circling the mountain. The water is put to order on the basis of "dragon shape" and formed into a system with pieces of dragon scales on the thin dragon spine, resulting in the most active and most pleasant waterfront. The meandering mountain forms the "dragon shape", yet the mountain peak is not on the extension of the central line so that the Jundu mountain chain on the north of Beijing can naturally merge into the forest park, the central line and the city.

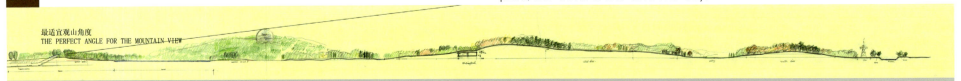

最适宜观山角度
THE PERFECT ANGLE FOR THE MOUNTAIN VIEW

二龙戏珠规划结构分析图
Double—dragon Layout Assay Chart

RHYTHMIC DANCE

规划的北中轴有南北、东西两种方向与节奏。南北方向延续历史轴线之周正，并有灵动的水线与之相伴，暗合了传统轴线之处理。轴线上并不置建筑，仅以平整的绿草为轴心且略加抬升，以东西向排列的繁茂乔木为骨架，由严整的序列渐渐隐入自然山水之中，既为中轴结束之余韵，又呈中轴尽端之绝响。东西方向为活泼空间与严肃空间之交替，有规整的树阵，弯曲的水系和平台、台地、坡地的变化。空间处理充分展示了城市轴线的时代特点，树、草、水穿插、完美的结合，为城市提供了一个富有纪念性的、充满寓意的、方便实用的舞台。

The planned northern axle line has two directions (south-north and east-west) and rhythms. The south-north direction extends the historical axle line proper and is accompanied by flexible waterlines, coinciding with the treatment of the traditional axle line. No structure is located on the axle line. Its axle center is made up only of even green grass, which is raised up a little. Its skeleton is the luxuriant arbor arranged in a southwest direction. The axle line disappears into the natural landscape bit by bit in a neat order. It is the aftertaste winding after the axle line as well as the last music sounding at the end of the middle axle. The east-west direction is alternative with lively space and serious space. In this direction, there are orderly tree arrays, curved water systems and a variance of flat land, terraced land and sloping land. The treatment of space sufficiently demonstrates the characteristics of the times of the municipal axle line. The perfect alternation and combination of tree, grass and water provides the city with a commemoratory, allegoric and practical stage.

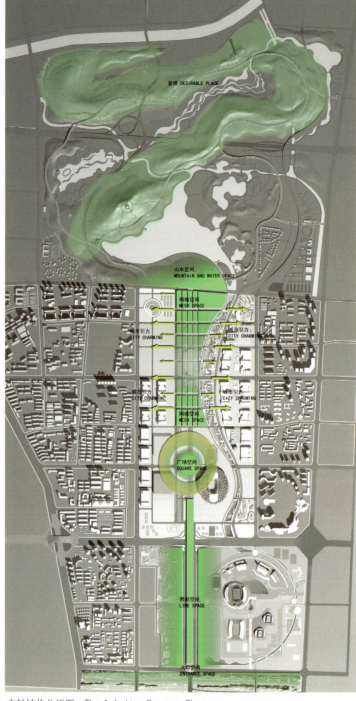

中轴结构分析图　The Axle Line Structure Chart

北京的传统轴线，皆为实轴。既内城、外城、皇城、紫禁城内的重要建筑多在轴线上设置，可见轴线的重要。轴线于紫禁城内所营造的空间氛围，似一阵紧似一阵的锣鼓点，给人以紧张、压抑与威严的感受；于皇城内所营造的空间氛围，则舒缓很多，体现了皇家的骄奢；于内外城所营造的空间氛围，就更加散漫，与城市的商店、水道、车道等相联系，世俗化了许多。

The traditional axle line of Beijing is real axis, i.e. the important structures inside the internal city, the external city, the royal city and the Forbidden City are mostly located on the axle line, from which we can perceive its importance. The space atmosphere built up by the axle line inside the Forbidden City is like the ever-intensifying gong and drum sound, giving people the feeling of tension, repression and majesty. The space atmosphere built up inside the royal city is much more slow and leisurely, reflecting the royal luxury and extravagance. The space atmosphere built up inside the internal and external cities is all the more loose and scattered, connecting with the stores, watercourses and roadways of the city, which makes the space more worldly.

森林公园植物种植规划图
Plant Planning Chart

花开花落，树荣树枯虽为自然规律，而古人却将四季的变化映射人生的无常。固有"春见山容，夏见山气，秋见山情，冬见山骨"之辩，也有"落花有意，流水无情"之说。连圣人孔丘也在水边感慨时间的流逝。"子在川上曰，逝者如斯夫。"

古今中外莫不把景物的四季不同作为造园置景的要素，加以发扬光大。分明的四季是北京最大的气候特点，北京人也最能体会四季给我们带来的欢娱。颐和园玉兰堂前春天的玉兰使我们温暖，北海公园里夏天的荷花让我们清爽，香山公园秋天的红叶体现的是热烈，天坛公园里冬天的古柏让我们凝思。此景此情，表达了我们对北京的热爱。

Although the blossom and fall of flowers and the flourish and withering of trees is the order of nature, the ancients hint at the uncertainty of life by the variance in the four seasons. So there is the argument that "In spring, you can view the appearance of the mountain, in summer you can sense the vital energy of the mountain, in autumn you can feel the passion of the mountain and in winter you can perceive the spirit of the mountain". There is also the saying that "The waterside flower pines for love while the heartless brook babbles on". Even the sage Confucius sighed with emotion at the lapse of time by the waterside, "It was by a stream that the Master said: Thus do things flow away, day and night".

At all times and all over the world, people have always taken the different scenes of the four seasons as the elements of gardening and scenery arrangement and have carried this tradition forward. The distinctiveness of the four seasons is the major feature of the climate in Beijing and the people here can experience most the great pleasure brought to us by the four seasons. The Yulan magnolia in front of the Yulan Hall of the Summer Palace gives us warmth in spring. The lotus in the sea of the Beihai Park brings us coolness in summer. The red leaves in Xiangshan Park make us enthusiastic in autumn and the ancient cypress in the Temple of Heaven lets us think deeply in winter. Such scenes and such passions express our deep love for Beijing.

四季之舞
Dance of the Four Seasons

Dance on the Journey

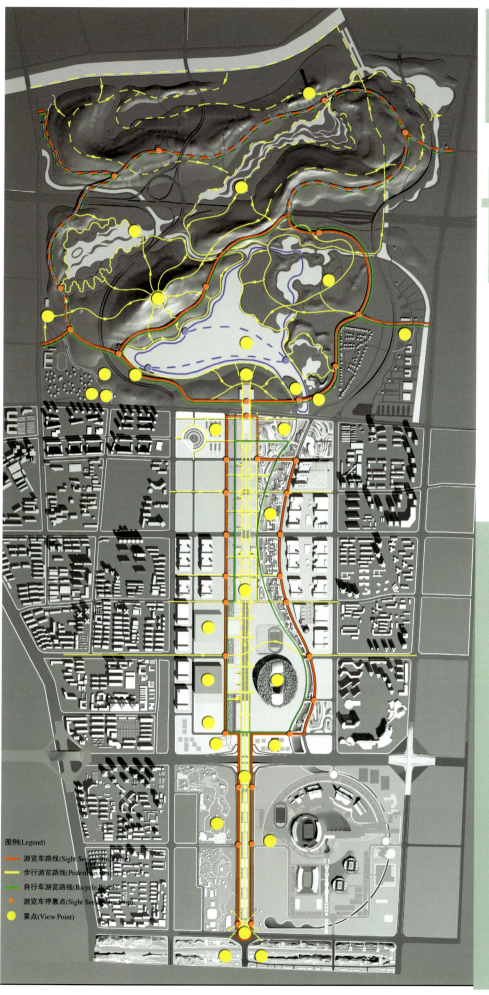

游览路线分析图
Sight Seeing Route Analysis

森林公园及中心区生态环境保护规划示意图
Eco-environmental Protection Planning Sketch

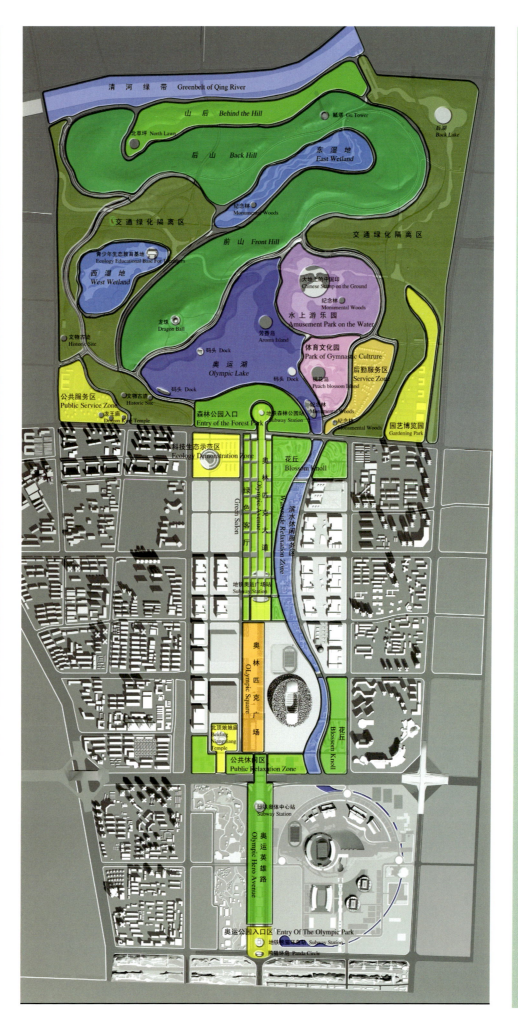

森林公园及中心区景观规划分析图
Landscape Planning Assay Map

Dance of Dragon

湿地景观效果
THE WETLAND PERSPECTIVE

湿地系统规划

规划根据森林公园景观规划所造成的地势，形成三块湿地，总面积约为 33.6hm²，通过该三处湿地来净化公园内河湖水系的水质，蓄滞洪水，蓄积公园内的雨水，并营造一种适宜动、植物生存的生态环境。

由于湿地植物在净化过程中起到十分重要的作用，因此选择合适的湿地植物非常重要。几个主要原则为：①耐污能力强和抗寒能力强；②选择在本地适应性好的植物，最好是本地植物；③根系发达，生物量大；④抗病虫害能力强；⑤最好有广泛的用途或经济价值。

Wetland System Planning

The surface relief created by the forest park landscape plan will form three wetlands, the total area being about 33.6 hectares. These three wetlands will purify the water quality of the river and lake water system inside the park, store and detain flood, accumulate and admit the rainwater inside the park and build up a sort of ecological environment favorable to the survival of animals and plants.

Since the wetland vegetation plays a very important role in the coures of purification, it is essential to select suitable wetland vegetation. The main principles of selection are: ① strong durability against pollution and strong chill-proof ability; ② Vegetaion with good local adaptability will be selected. It is preferred that local vegetation is selected; ③ Flourishing root system and large biological mass; ④ strong ability against plant diseases and insect pests; ⑤ they might as well have extensive usage or economic value.

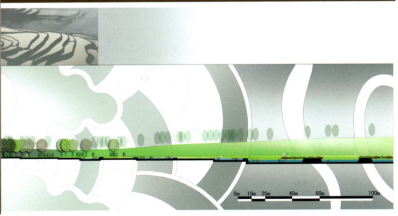

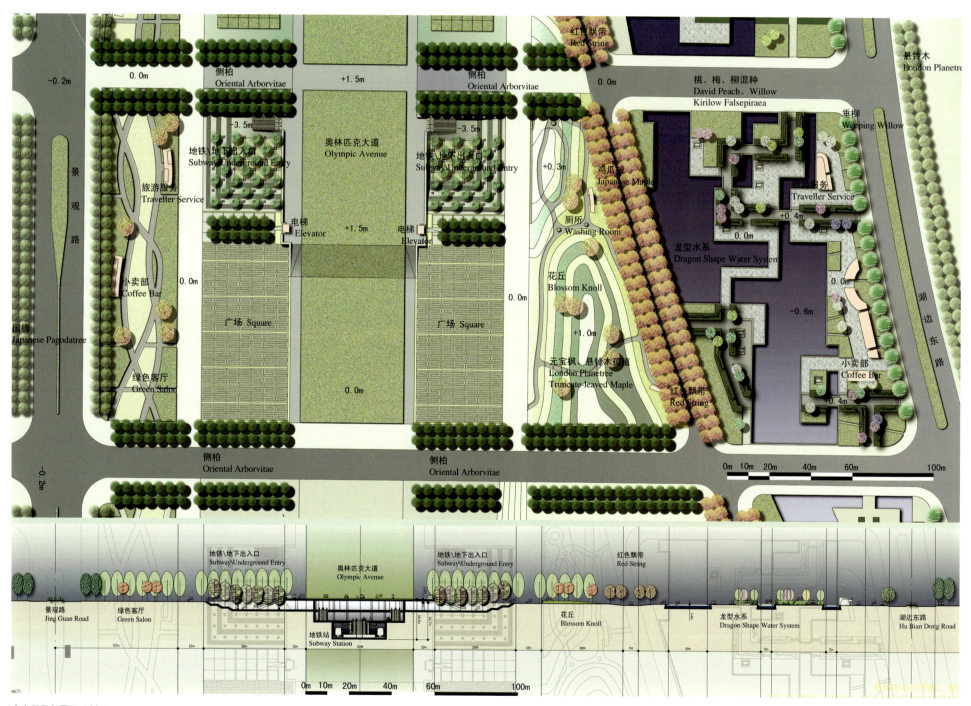

中心区局部平面／剖面
The Part of Central Zone Plan/Section

DANCE OF DRAGON

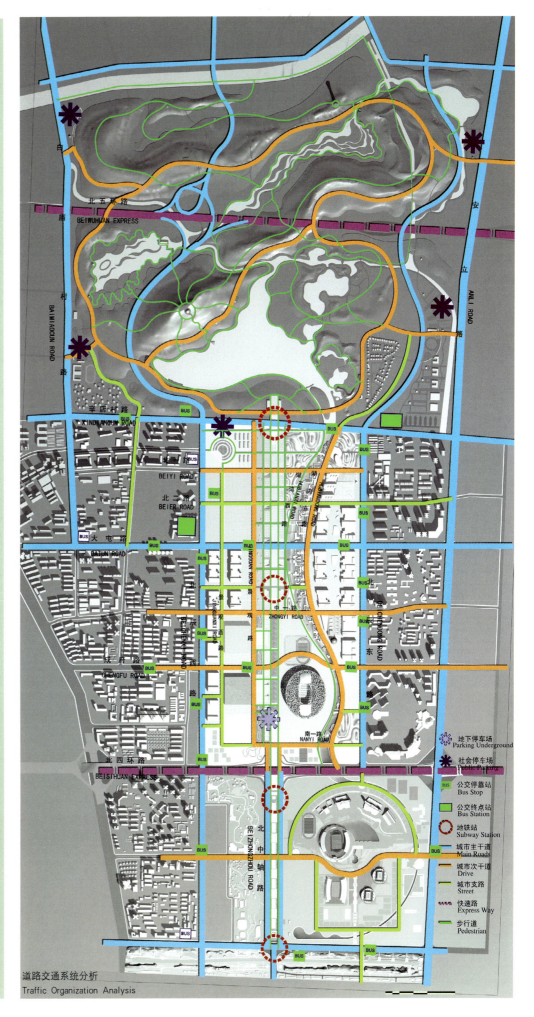

道路交通系统分析
Traffic Organization Analysis

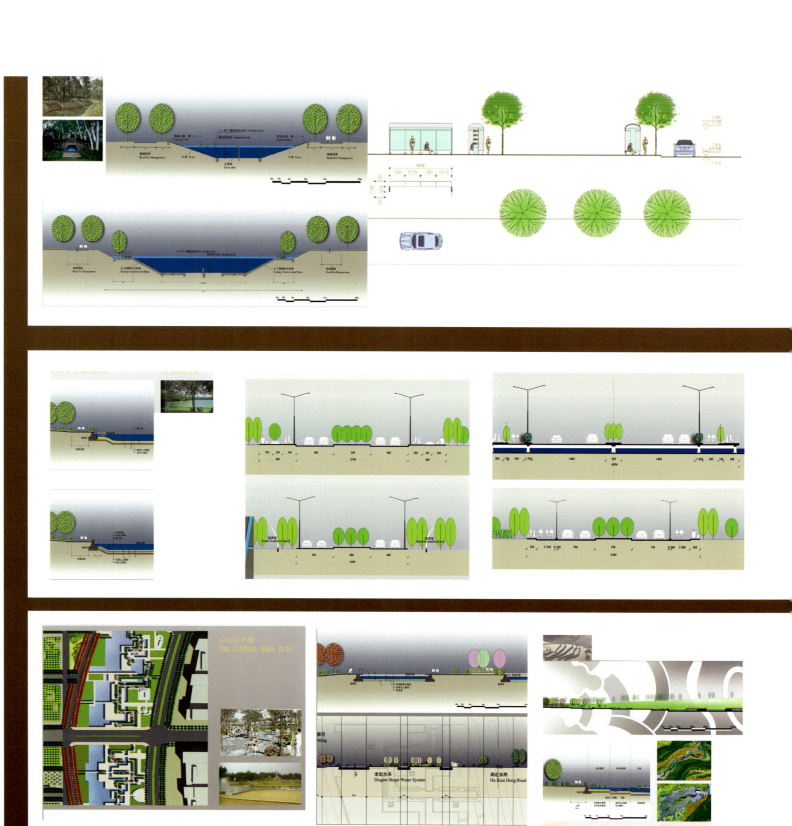

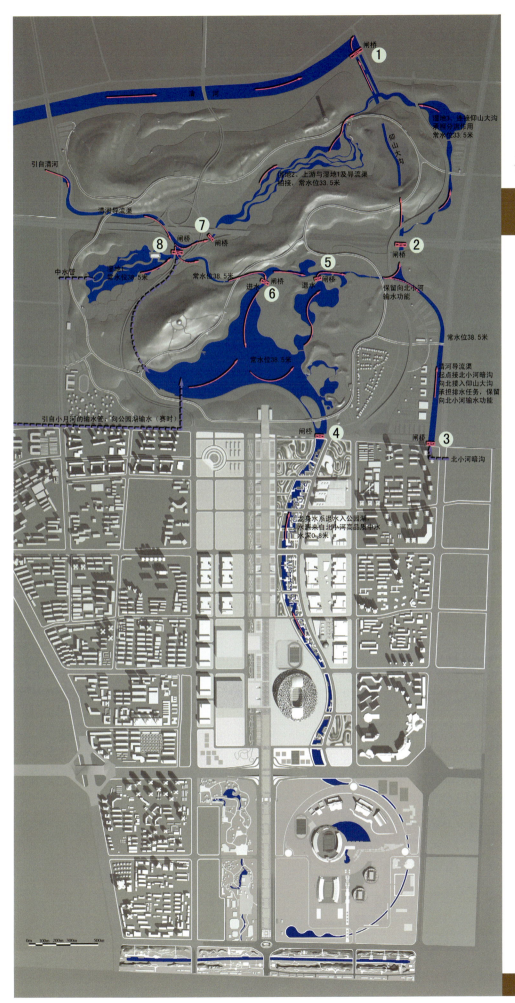

水系规划图
Water Planning

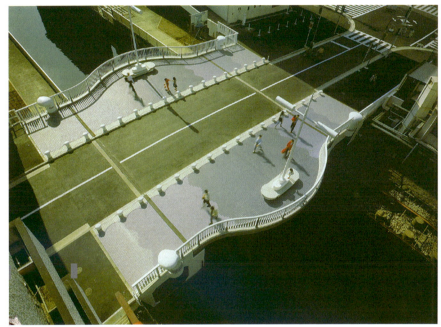

A04
Turen Design Institute

北京土人景观规划设计研究所

奥林匹克公园森林公园及中心区景观规划设计方案

OLYMPIC GREEN LANDSCAPE DESIGN FOREST PARK AND CENTRAL ZONE

田

　　五千年的中国土地曾经、还将养育世界上最多的人口。在无数的失败与成功之后，田，积淀下了处理人与土地关系的最精华的智慧。连同那充满诗意的景观，田，告诉我们如何尊重土地和自然过程，用最少的工程，获得可持续的最大收益。在快速的城市化进程中，这五千年智慧连同其诗意的景观被无情地毁弃，我们甚至以"工业化"和"高科技"的名义来"整治"和"改造"我们生命的土地、水系，而最终发现我们祖先对待土地的态度和技术要高明的多。

　　当城市化和可持续挑战中国大地，奥运会成为未来北京的主题时，奥林匹克公园的设计应采撷五千年的造田、种田、灌田的技术、艺术、和精神，并将其与现代最新生物和能源科技相结合，实现一个科技的、人文的、绿色的奥运景观典范。一个未来中国和世界的可持续景观典范。

Memory and Prophecy

The land of china with its five thousands years of agriculture history has been worked and reworked. The footprint of the past is in her terraced farmlands, terraced waters and mountains, and sustainable agrarian/ water practices. This footprint integrated with cutting-edge biological and energy technologies forms the foundation of the design of this historic Olympic Forest and Axis. These elements are combined into a 21st century sustainable urban landscape that embodies the spirit of the past-memory and speaks to the future, Prophecy.

In an era when China is undergoing an unprecedented process of reinventing herself as an urban, 21st century country, the Beijing 2008 Olympic games is a direct expression of this process. The most important mission for the Olympic Park is to combine the knowledge from the past with the future to create a model of a sustainable landscape. The park is an expression of the timeless power of this land to transform.

记忆 / MEMORY

造田 Field Making

种田 Field Planting

127

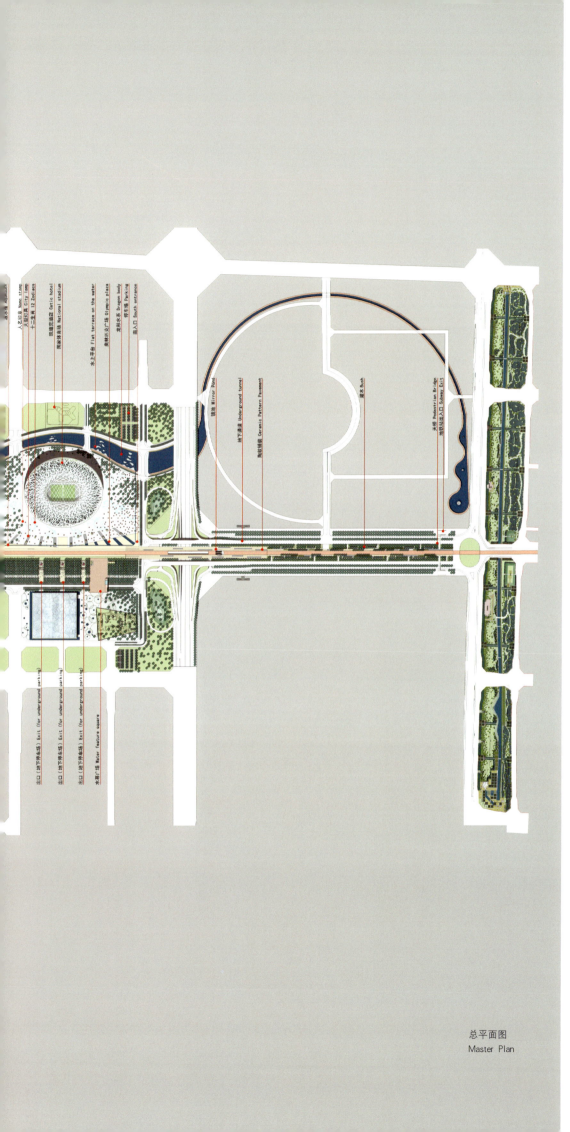

总平面图
Master Plan

田的纪念 Field Monumentality

田的欢乐 Field Celebration

129

灌田 Field Irrigating

田的欢乐 Field Celebration

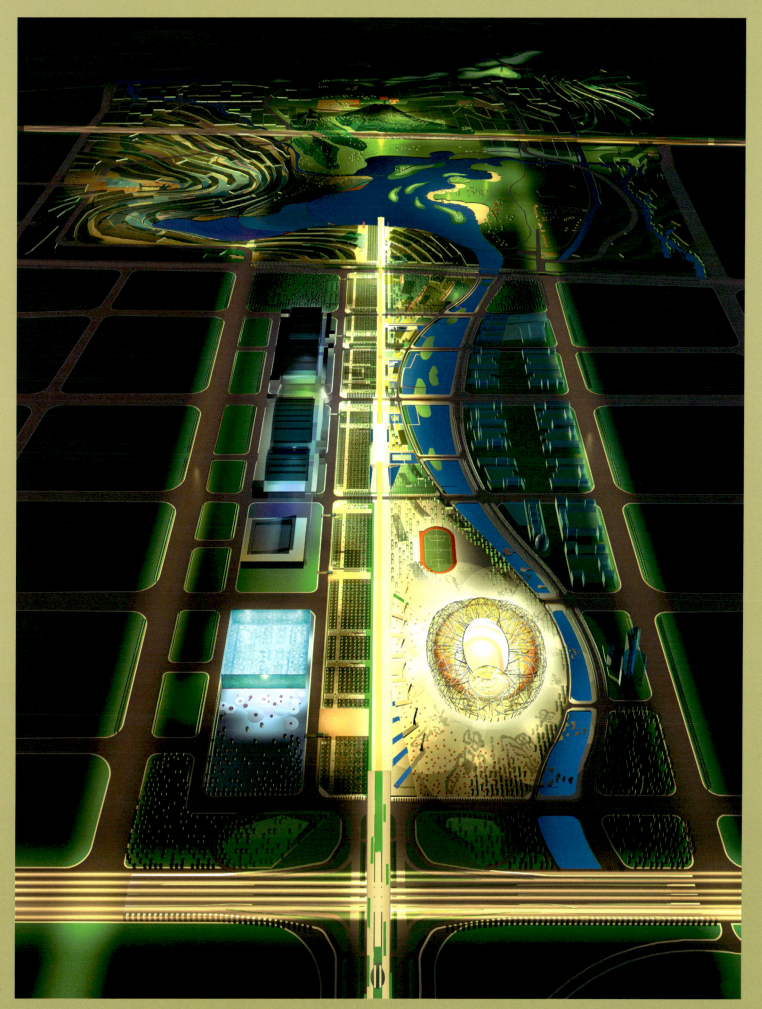

总体鸟瞰图
Bird-eye View

可持续的水系统：渡水槽和水体净化、循环和再用
A Sustainable Water System and Aqueduct: reduce, reuse and recycle

轴线：五线谱上跳动的音符，一部关于五色土地的交响乐
The Axis: a symphony of five landscape elements

场地故事：环境解说系统与专类植物园
Stories from the Site: story lines and special botanical gardens

六大设计特点
Six Major Design Solutions

梯级湖群+梯田湿地：最少工程量的场地设计途径
Terraced Lakes + Terraced Wetlands: a minimum approach

田+等高线：结构与象征性肌理
"Contour + Fields": the structural and symbolic fabric

植被：景观生态模式与"三皱法"配置
Vegetation: the "Matrix-Patch-Corridor" model of landscape ecology

133

湿地与梯级湖群
Wetland and Terraced Lake

Memory and Prophecy

鸟瞰图
Bird-eye View

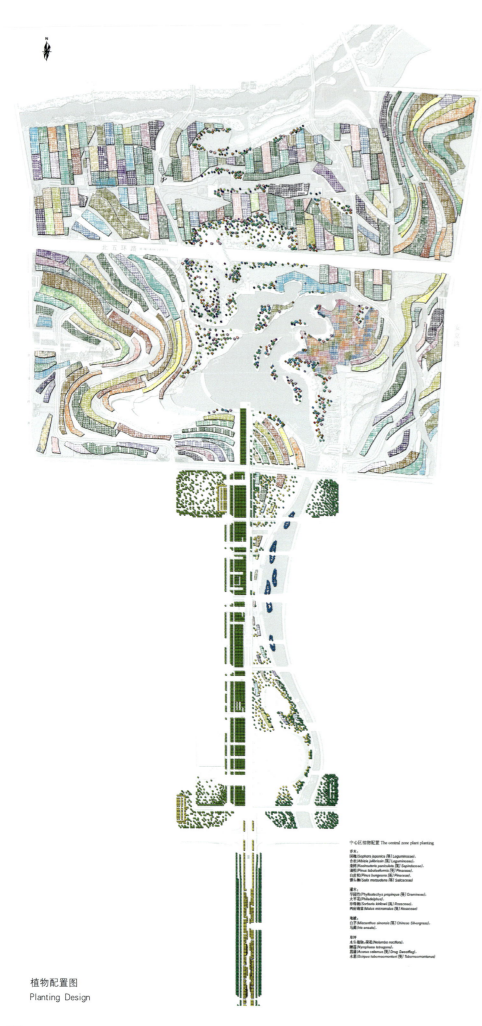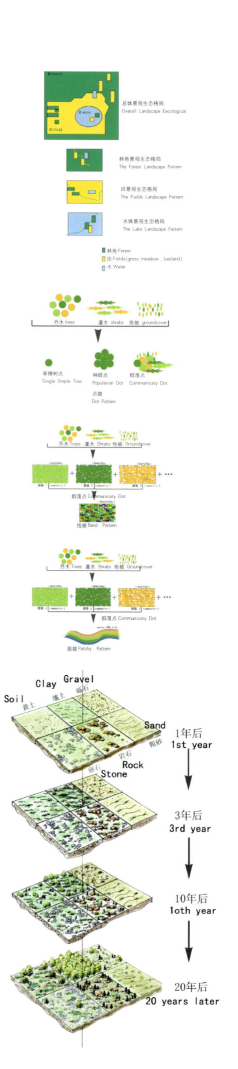

植物配置图
Planting Design

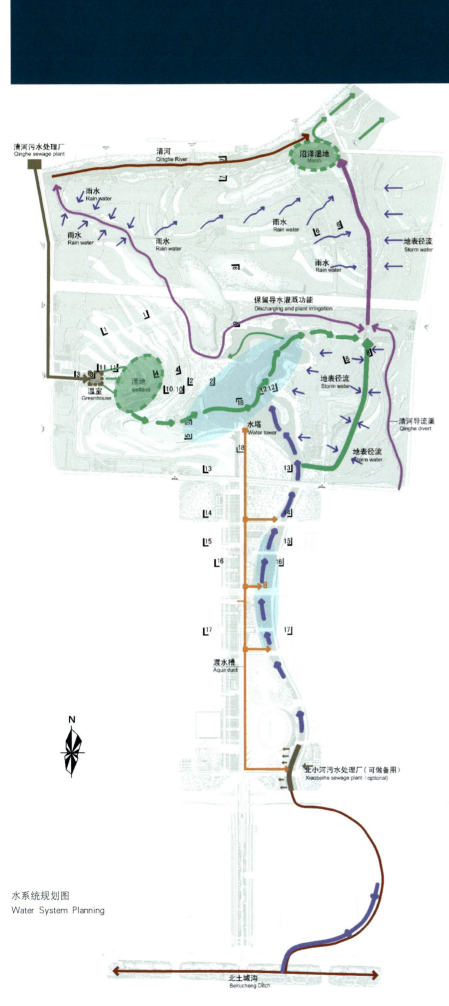

水系统规划图
Water System Planning

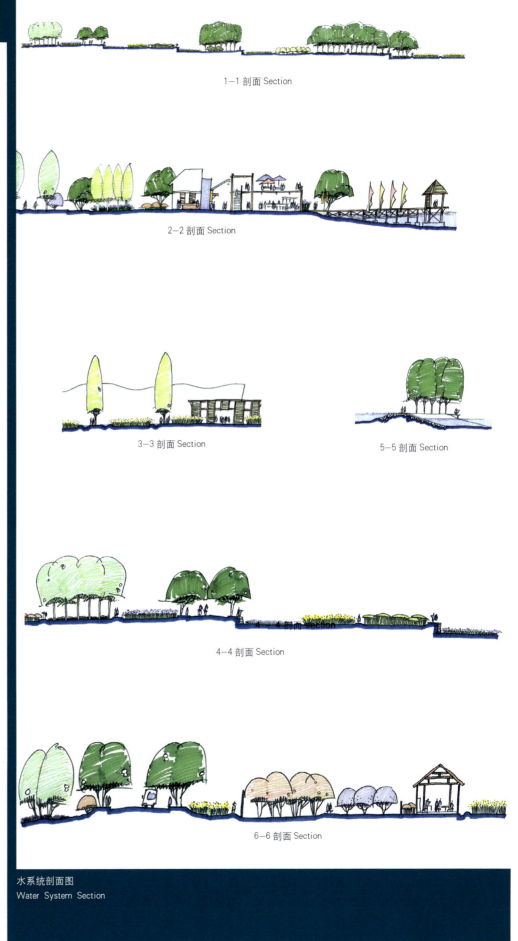

1-1 剖面 Section

2-2 剖面 Section

3-3 剖面 Section

5-5 剖面 Section

4-4 剖面 Section

6-6 剖面 Section

水系统剖面图
Water System Section

水系统剖面图

Water System Section

7—7 剖面 Section

8—8 剖面 Section

9—9 剖面 Section

10—10 剖面 Section

11—11 剖面 Section

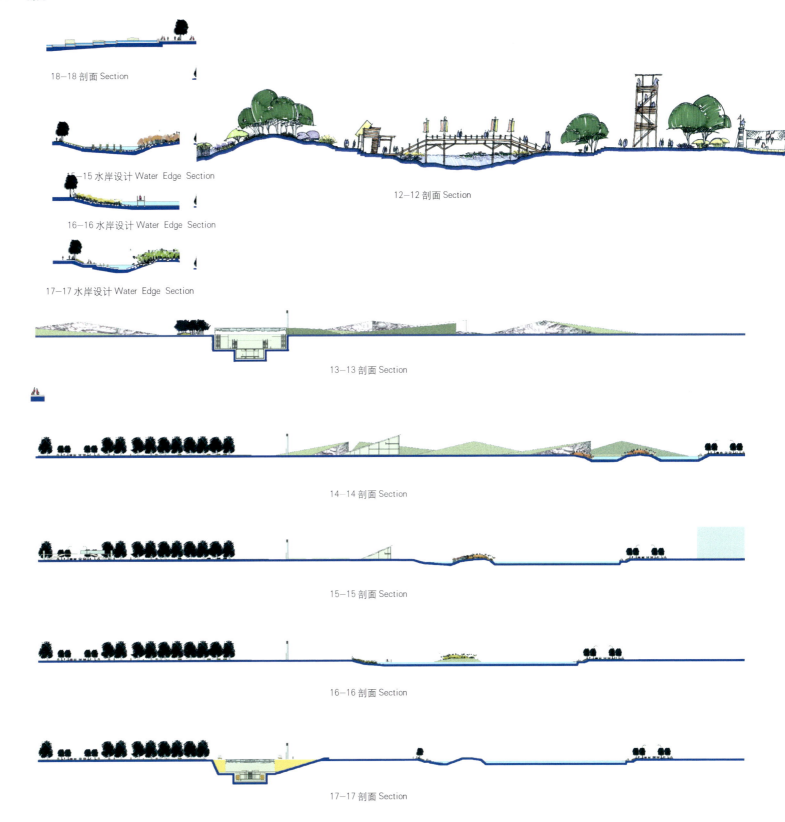

18—18 剖面 Section

15—15 水岸设计 Water Edge Section

16—16 水岸设计 Water Edge Section

17—17 水岸设计 Water Edge Section

12—12 剖面 Section

13—13 剖面 Section

14—14 剖面 Section

15—15 剖面 Section

16—16 剖面 Section

17—17 剖面 Section

OLYMPIC GREEN LANDSCAPE DESIGN FOREST PARK AND CENTRAL ZONE

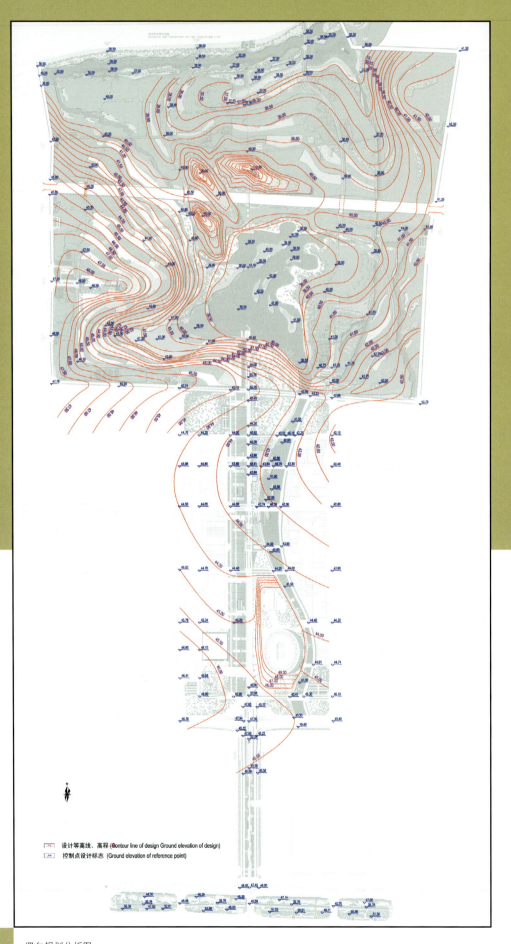

竖向规划分析图
Grading Design

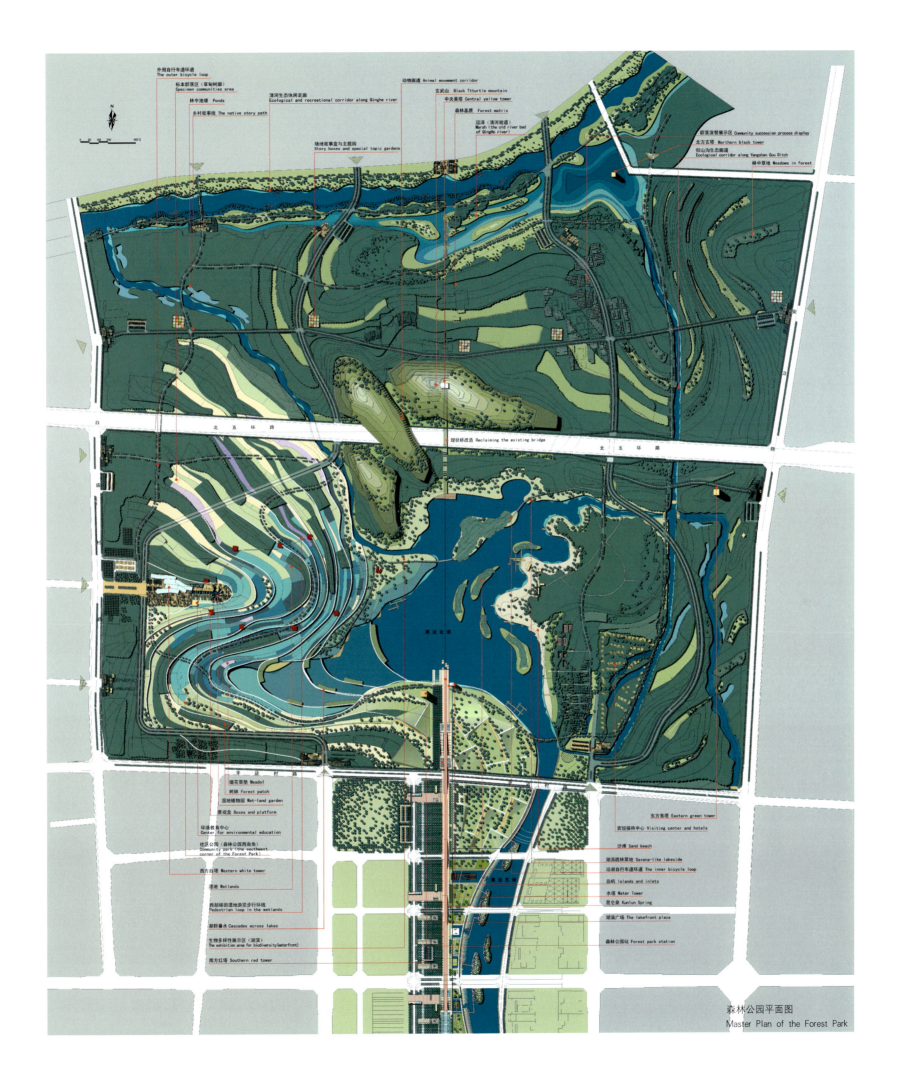

森林公园平面图
Master Plan of the Forest Park

森林公园透视效果图
Forest Park Perspective

Memory and Prophecy

生态农庄（温室前）
Eco-farm Planting (in front of the green house)

指状园林（标本群落展示）
Forest Fingers on the Meadow

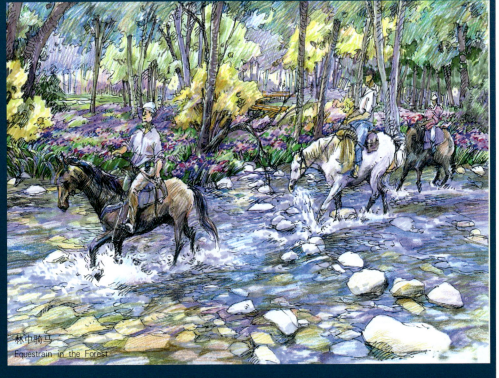

林中骑马
Equestrain in the Forest

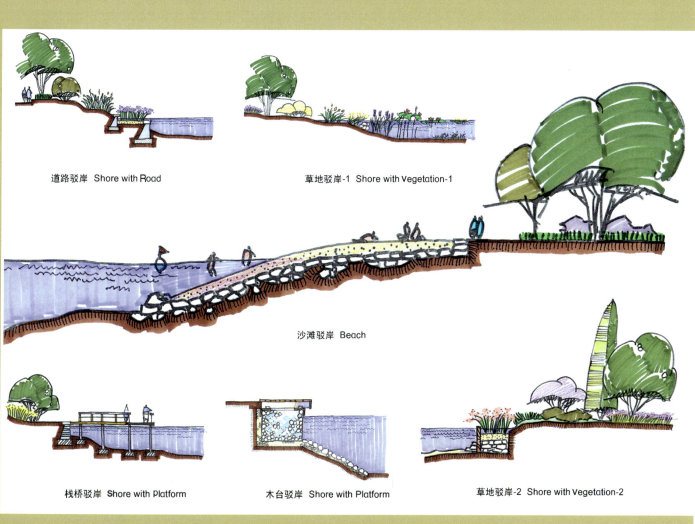

道路驳岸 Shore with Road　　草地驳岸-1 Shore with Vegetation-1

沙滩驳岸 Beach

栈桥驳岸 Shore with Platform　　木台驳岸 Shore with Platform　　草地驳岸-2 Shore with Vegetation-2

水岸设计
Water Edge Design

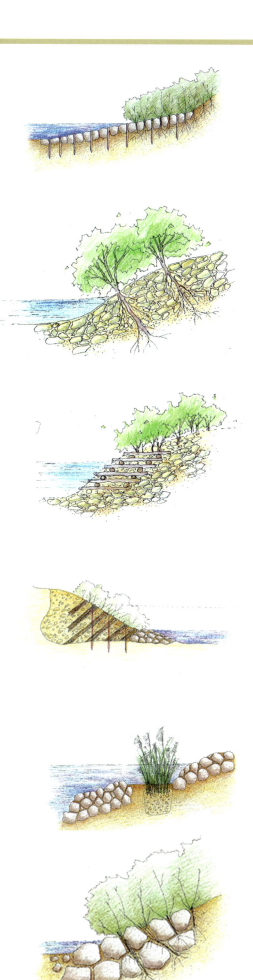

森林公园分析图
Porest Park Analyzing Dilagram

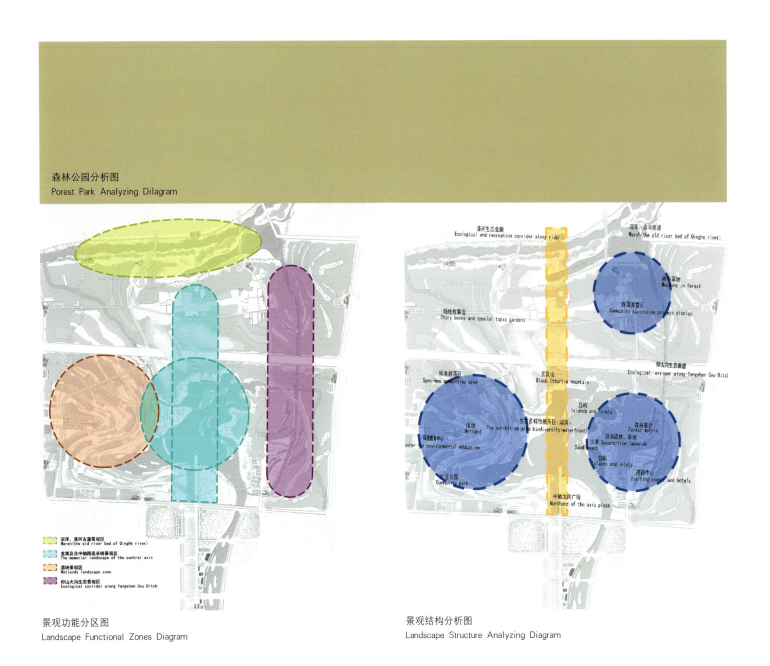

景观功能分区图
Landscape Functional Zones Diagram

景观结构分析图
Landscape Structure Analyzing Diagram

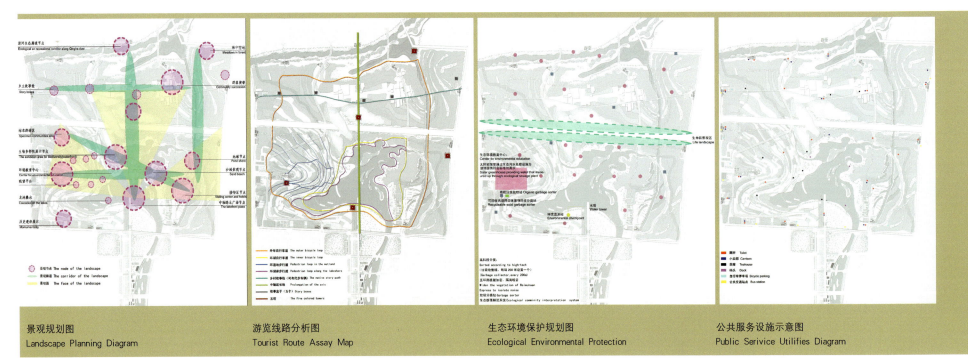

景观规划图
Landscape Planning Diagram

游览线路分析图
Tourist Route Assay Map

生态环境保护规划图
Ecological Environmental Protection

公共服务设施示意图
Public Serivice Utilifies Diagram

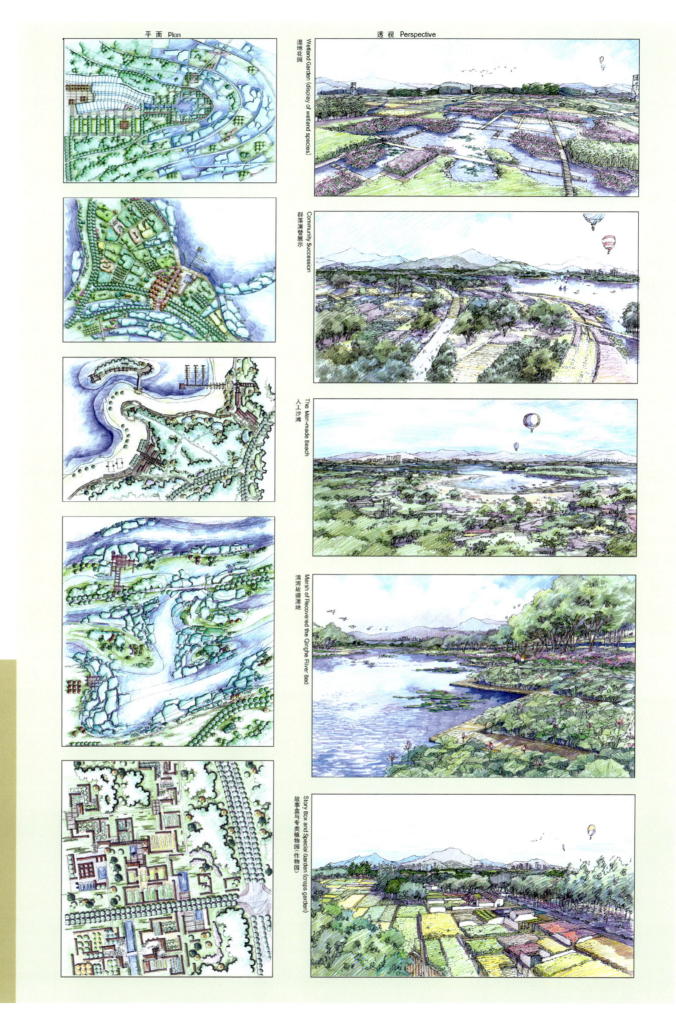

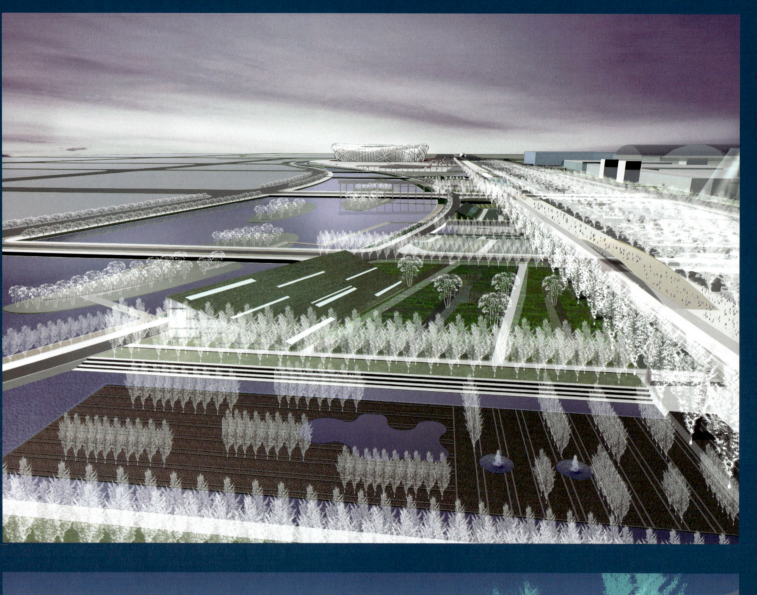

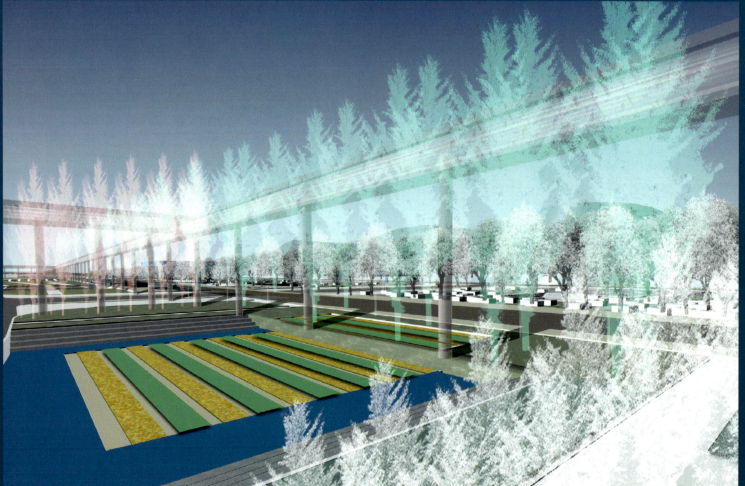

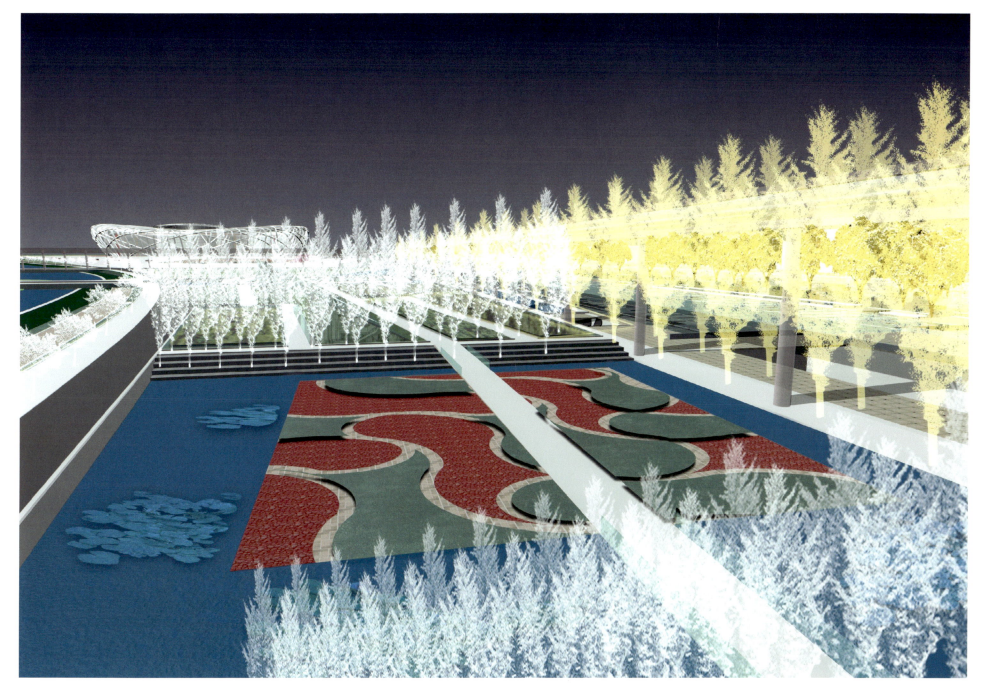

Memory and Prophecy

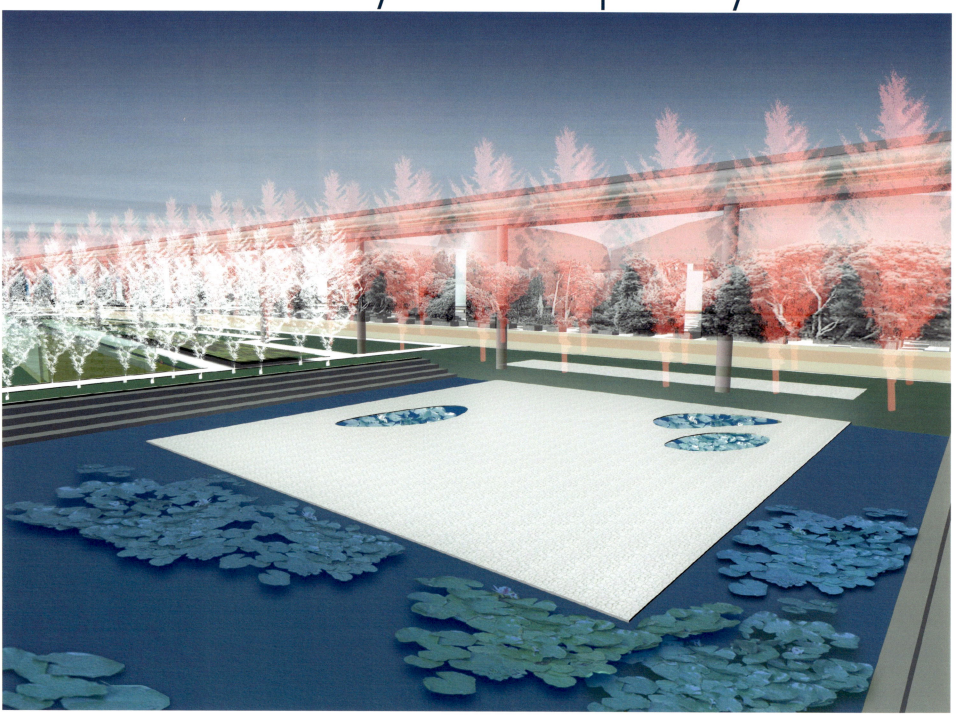

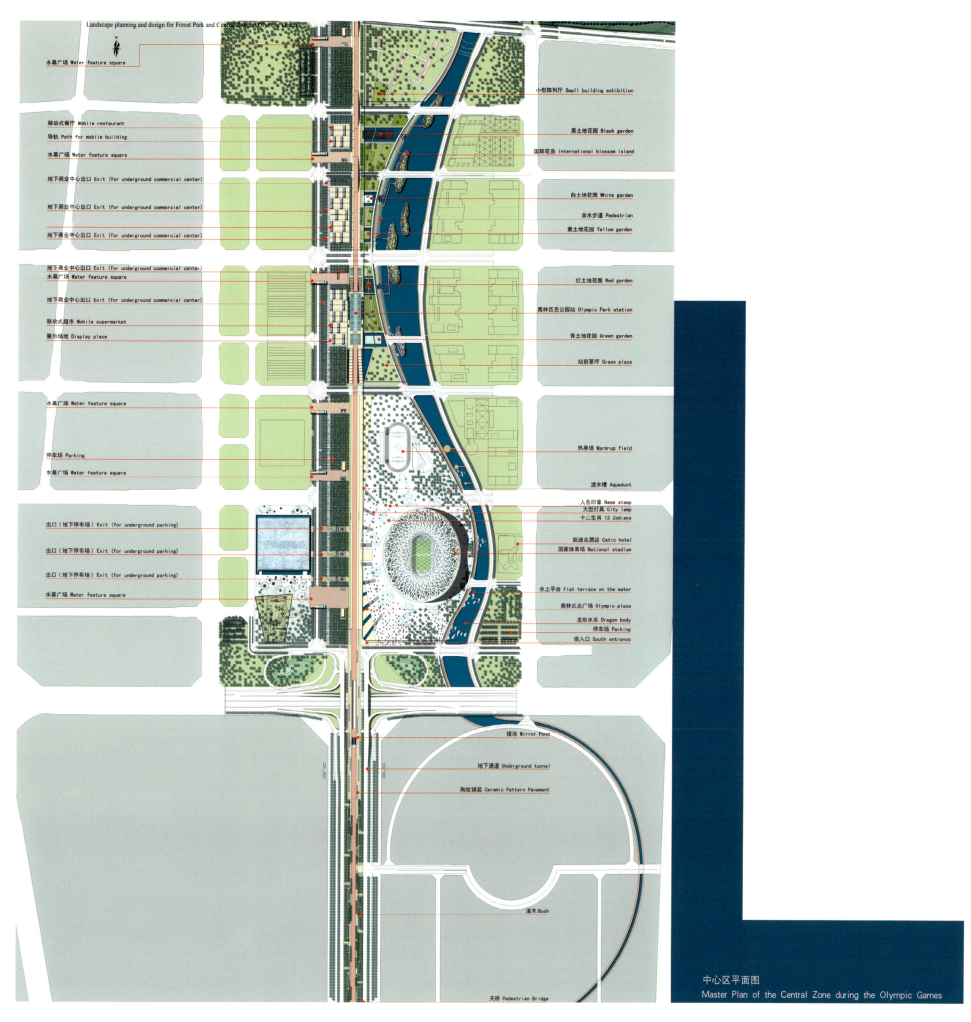

中心区平面图
Master Plan of the Central Zone during the Olympic Games

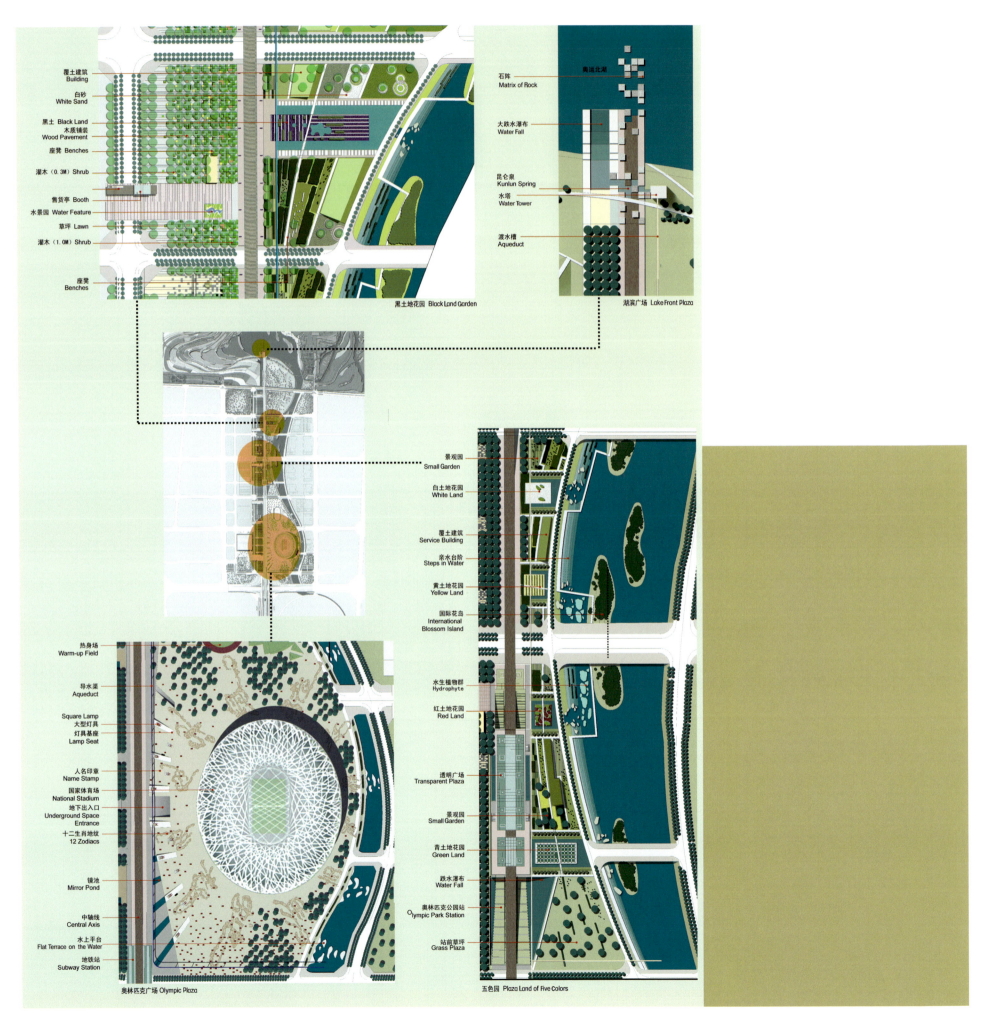

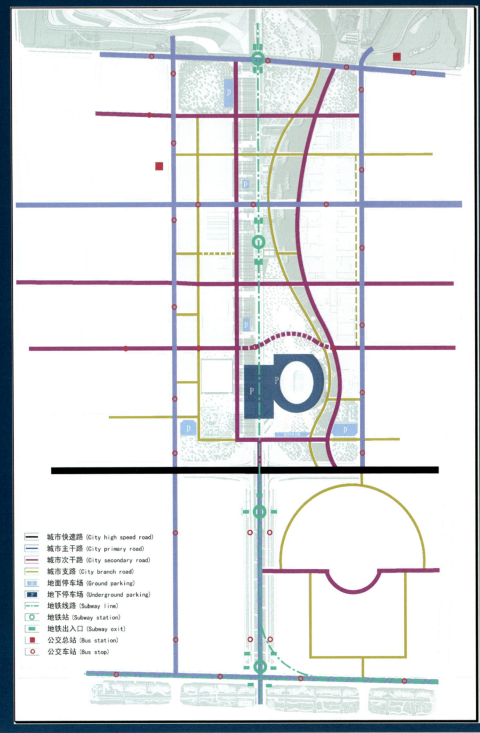

交通系统规划分析图
Traffic Planning Chart

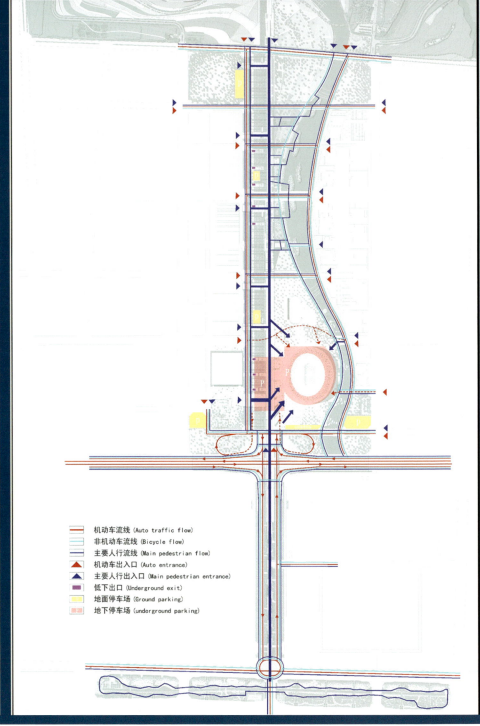

道路交通组织规划分析图
Road and Flow System

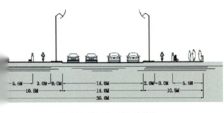

南一路、中一路、北一路断面
Nanyi Rode/zhongyi Rode/beiyi Rode Section

湖边西路断面 Hubian Xi Rode Section

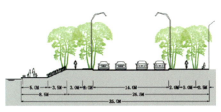

湖边东路断面 Hubian Dong Rode Section

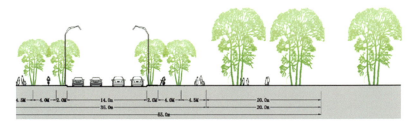

景观路断面 Jingguan Rode Section

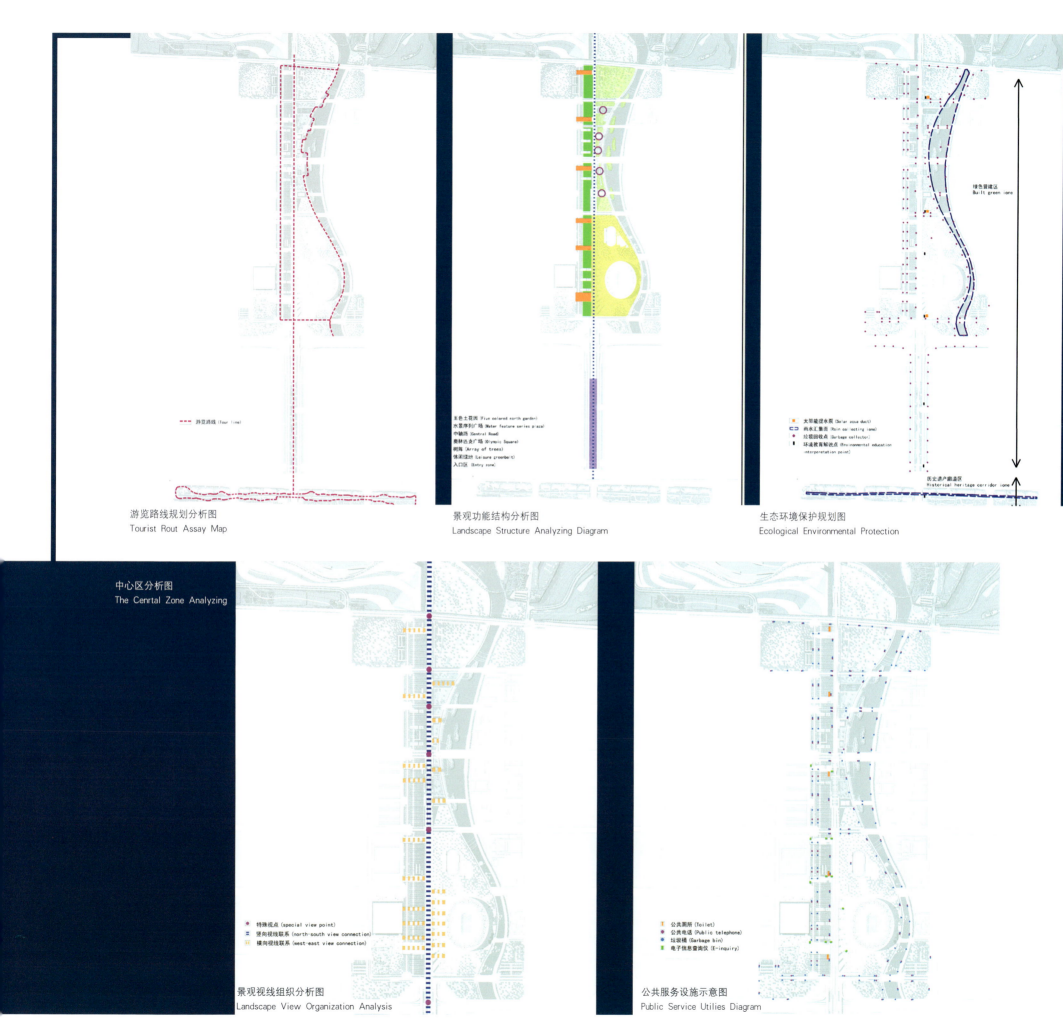

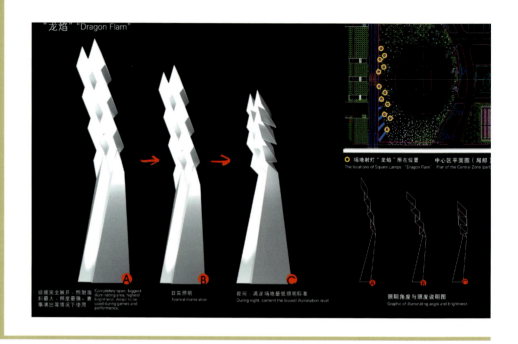
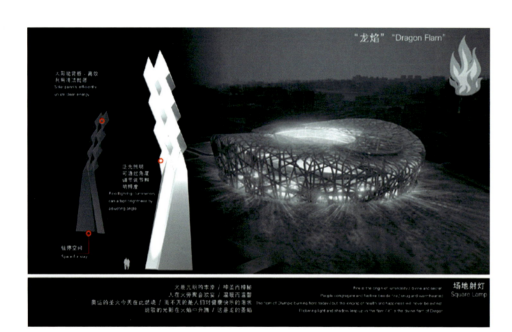
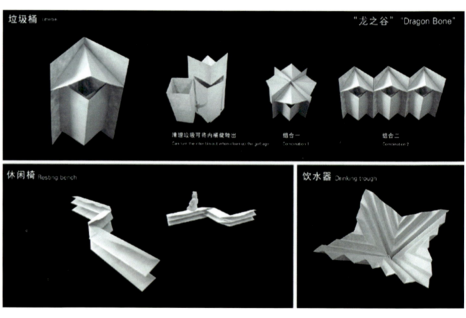

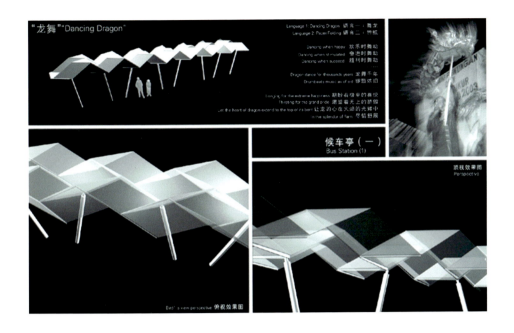

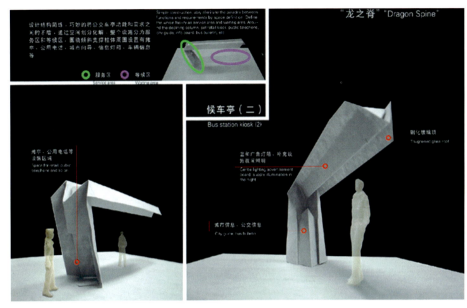

赛时的赞助商展台、快餐厅、超市等服务性建筑为装配式轻钢结构，下装滑轮可在铺设于场地中的轨道上滑动。移动式的建筑可充分适应不同的展商对展场大小的需求及不同时段对场地使用性质的不同需求。同时艺术装置般的移动建筑构成了场地中不断变化的风景。
The service buildings such as the displaying space for the sponsors, retailing and snack are designed that can be moved along the path in order to meet the variety needs of sponsors and customers. In the meantime, the moving device shape a variety landscape.

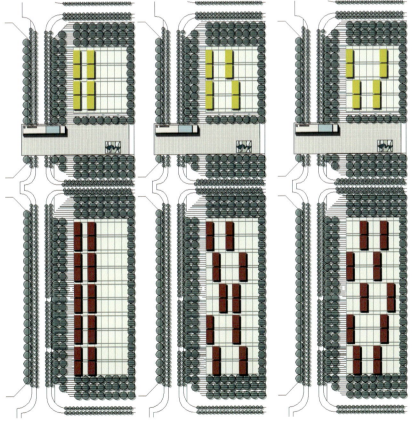

1. 非展出时间，提供更多的行人空间
at the time of non-exhibition, the space is main for the pedestrian

2. 展出时间，提供更多的展示空间
at the time of exhibition, there's more space for displaying of sponsors

3. 随着赞助商的变化，展台的布置及展示场地的大小可以灵活地适应
the distribution of stage can change with the variety needs of sponsors and customers

A05

瑞驰·汉格及合作者公司
Rich Haag & Associates

XWHO(Hangzhou) Inc.
艾斯弧（杭州）建筑规划设计咨询公司

奥林匹克公园森林公园及中心区景观规划设计方案
OLYMPIC GREEN LANDSCAPE DESIGN FOREST PARK AND CENTRAL ZONE

鸟瞰图
Bird-eye View

SUMMARY 概 述

北京森林公园

在生动活泼且有序安排的中心区北部，不规则自由设计的森林公园的中心部分是一个龙形水体的大湖。沉湖四周，大片柔软的草坪，和大片景观造型，疏密有致种植的植物共同定义了各种不同的休息、娱乐运动空间。

等级层次分明的道路系统，满足不同空间之间的联系，同时保证满足园内服务要求。五环路两边被密集种植植物和地形隔离。在公路上架一个巨大的天桥欢迎游人深入到北京奥运森林公园的中心部分"萌丘"。萌丘，"MANATURE"，成为"人与自然和谐共有"的代名词。

北部区域分三部分：
农田景观区（人与自然的关系）
大地景观和活动区（人与人之间的关系）
原始生态区（自然不受人干扰）
生态湿地
保留实验地（没有人为干扰）

西部农田景观区是一片丰富多样的大地景观活园林，包括水田景观、湿地植物、高山农田、果园景观以及林业景观等，所有这些都体现了中国人农牧渔业的悠久历史，也表达了人与自然之间的和谐关系。

东部原始生态区域，表达了从时间历程排列上更为远古的历史文化的缩影。设计思想来源于生态学家的灵感，通过象征不同历史时期的时间长廊过渡到更野性和自然的原始生态区域，中心部分是一系列的湿地水塘，四周地形合理起伏变化，以适应各种花卉、花草更繁茂生长。

东北部的观察实验区是面积1hm^2的一片空地，作为一个开放的自然实验场，让人观察自然的进程如何使一片荒地重新恢复到生机活力的过程，体会自然最原始而伟大的力量。

农田景观及原始生态区通过渐变的纪念性大地造型空间联系起来，成为中轴线上阴阳平衡的两个区域。内部一个月牙型小岩洞，沿岩壁做成刻浮雕记忆之墙，它的体量设计成一个合适尺度的圆形剧场，可以用以举行仪式和庆祝活动。从两个高峰上可以眺望2008奥运场馆全景和森林公园全貌。

大地艺术景观的营造，将给北京森林公园带来一种恒久的影响力。从真实的场景体验看，以萌丘为中心的大地艺术部分和以"鸟巢"体育馆为代表的标志性建筑，一阴一阳的景观和建筑之间，将构成奥运公园一个和谐的整体。

Beijing Forest Park

North of the very active and programmed center zone the great lake is encircled with an informal park. Large meadows with soft earth forms and tree plantings define spaces for a diversity of recreational activities from organized sports fields, free play games, picnics, family reunion to quiet passive pursuits.

A system of earth forms, plantings and paths organize the wide range of social interaction and provides service functions. Dense planting on earth berms on both sides isolates the Wuhuand road. A broad overpass over the highway invites visitors to explore the heart of Beijing Olympic Forest Park.

The north sector has 3 zones:
The agrarian model (man with nature)
The earth form (man with man)
The primeval zone (nature without man)
Ecologically wetland intervention
Unplanned (no intervention)

The west: the agrarian model demonstrates historic Chinese dependence on the cultivation of food, a pattern, a living garden with rich diversity of forms aquaculture, wetland crops, upland fields, plantations, orchards and managed forests, all designed to symbolize the reciprocal partnership of man with nature.

The east: the primeval zone completes the prehistory of the MANATURE wheel. Through the passage of time, this idea (with a jump start from the deep ecologist) will mature into a metaphorical wilderness, a tribute to the indifference of untamed nature. The heart of this diverse environment is a series of deep-water ponds, with wetlands and an undulated topography all arranged to maximize the flora/ fauna evolution.

NE (north east) of primeval zone is one hectare of strong earth sculptures to be abandoned as raw and barren, this open laboratory demonstration of the self-healing power of natural processes to revivify a vacant wasteland.

Sectors 1 and 3 are united by ceremonial earth formations that are balanced on the axis of Beijing. The interior reveals a cavern form growing upward from a curved memorial wall. This formation is serves as a sound stage for ceremonial or festive performances projected across to the large amphitheatre. Panoramic views of the 2008 Olympic site and the countryside are offered from both summits.

The solid power of the earthwork gives authority and timelessness to the Beijing Forest Park. In a real sense, this feature is the earth sister to its brother symbol, the bird-nest stadium.

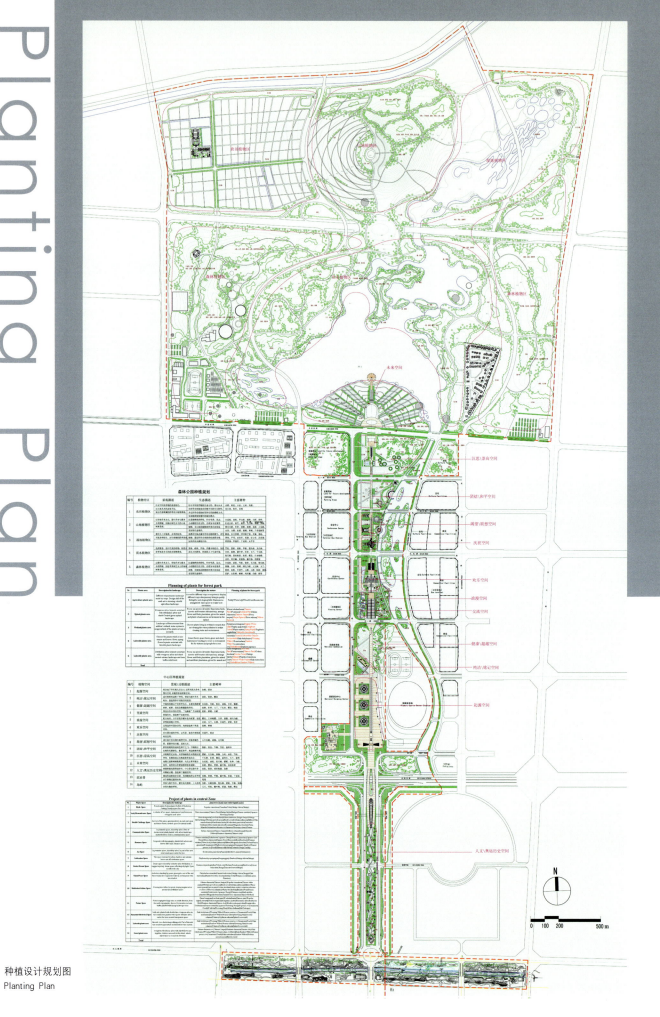

种植设计规划图
Planting Plan

总平面图
Master Plan

赛时规划总平面图
General Layout during the Olymic Games

Design Principle

在整个景观规划设计里，我们始终秉持奥运中心区和森林公园的景观应该成为陈述和表现"人文奥运，科技奥运，绿色奥运"精神的表现载体。因此，设计方案无论是对人文、文化的表达，生态景观的表述，还是高科技的运用和表现，都从多角度多层面精心组织，努力实现将设计从文化，生态，科技和艺术几方面之间和谐相融的效果，达到"设计源于生活，文化融入自然，科技回归人性"的最终目的。

设计方法
DESIGN METHOD

传统设计手法表达非传统特征元素(A)；反传统设计方法表达传统特征元素(B)。

The traditional design method demonstrates nontraditional characteristic elements; The anti-classic design method demonstrates traditional characteristic elements.

(A) 传统设计手法
Traditional design method

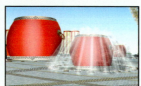

(B) 反传统设计方法
Anti-classic design method

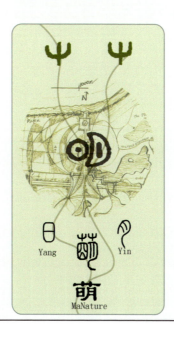

传统 v.s. 现代
Traditional v.s. Modern

软质 v.s. 硬质
Soft v.s. Hard

路：规则 v.s. 自由
Road: Order v.s. Free

植物种植：规则 v.s 自由/疏 v.s 密
Planting Style: Order v.s. Free;
Loose v.s. Dense

自由变化 v.s. 规则秩序
Free v.s. Order

生态自然 v.s. 人文文明
Natural v.s. Cultural

空间：复杂 v.s 简单变化
Space: Complex v.s. Simple

我们充分理解景观作为联系建筑物和空间之间的过渡和平衡作用，继承了中心区的"鸟巢"主场馆和"水立方"游泳馆设计对于中国传统"阴阳"哲学的体现。而且，对于"阴阳合，万物生"的中国哲学思想的诠释，也是我们对于"运动是生命的源泉"这一奥运精神和运动宗旨的最好表达。所以，在整个方案之中，无论在景观元素特征，设计表现方法等方面都充分体现了"阴阳"平衡的传统哲学思想。

The golden rule governing this design, is the application of the historic Chinese philosophy universally known as the yin and yang principle "yin and yang combine then create life,"

设计基本原则
Design Principle

■ 表达人文、文化的景观元素
Cultural Landscape Elements

我们强调"人文奥运"应该是中国文化和奥运文化的一次撞击和融合。因此，在景观设计上，我们将中国文化元素的体现充分反映到空间的各个层面。不仅仅使大众感受到奥运是"北京的奥运"，而且借助于奥运文化的传播，将中国文化和文明的讯息带给全人类。
In each of the theme spaces, the ancient Chinese culture informs the Olympics culture. The visitor (and the media) will convey to the whole world the image of the real Beijing style Olympic.

■ 文化和文明跨越时空的交流和对话
Different culture Dialogue

设计通过对景观元素"时空转换"的处理，将景观的正常时间元素进行错位，产生文化跨越时空对话和交流的效果，并且隐喻中国文化和奥运文化以及世界文明之间潜在的联系。
The beginning of first kilometer central axis: a space/time line separates yet couples Chinese civilization with Olympics Culture.

■ 设计对于科技的阐述
Design for interpreting and integrating technology

1、高科技信息服务设施
Facilities of High-tech information service

在景观设计里，预留了足够的空间以备建奥运会期间的高科技信息服务设施，包括虚拟观赛平台，赛事电子记录屏，互动触摸屏信息查询点，等等。
In landscape design, reserve the enough space to have high-tech information service facilities, include the conjecture view match terrace, the game electronics record holds, interacting the touch hold the information search orders, etc. That set up the period of Olympic game.

2、高科技材料的运用
High-tech material and methods

奥运"火炬长城"：高科技材料的运用，使真实的火和模拟的火景共同形成了一条"火炬长城"。
A skillful combination of real fire, pyrotechnics and simulated fire is programmed to celebrate small daily events to the finale a progressive procession of fire flowing up to the axis, climaxing in a galactic fire storm over the great mother earth form.
采用特殊材料制作的"奥运记录墙"
Adapt modern materials and techniques to create the Olympics record wall.

3、高科技文化的表达
设计对科技的表达为了普及和教育作用
Landscape pattern inspired by interpretation of high-tech sources X-Y Chromosome, computer chip etc. to explain, to educate the general public.

交流空间：交流空间的景观空间纹理的设计构思来源于电脑的芯片纹理；以此来反映信息时代的代表性特征。设计里，流动的水"模拟"信息的传播，而树和人的动态空间成为寓意"信息交流"交换的节点，表达了信息时代信息交流产生生命的含义，也同时反映了交流空间的主题。
This landscape design is derived from magnified form of computer chip. Flowing water signifies the flow of electrons, and the movement of information through the circuitry. Dynamic relation of man to the tree of knowledge is another metaphor for the many emerging branches of intercommunication.

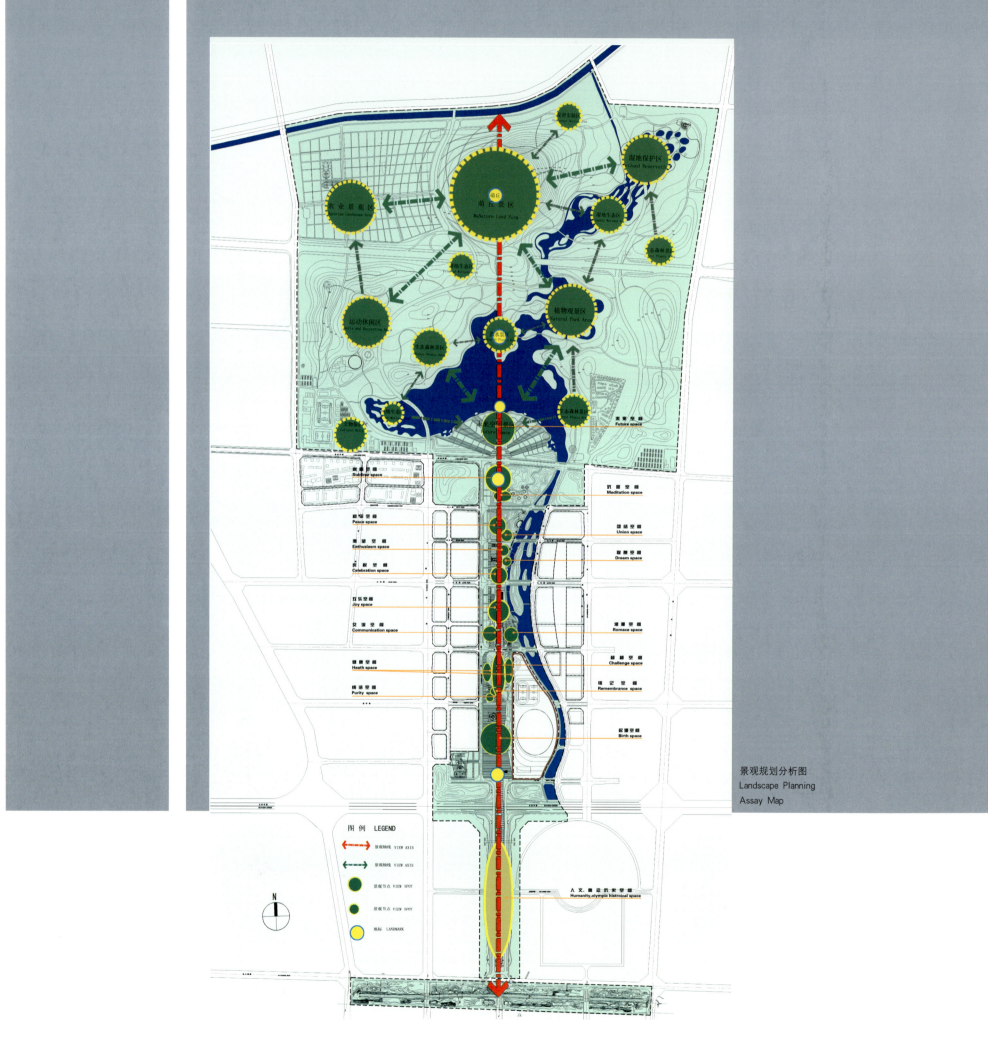

景观规划分析图
Landscape Planning Assay Map

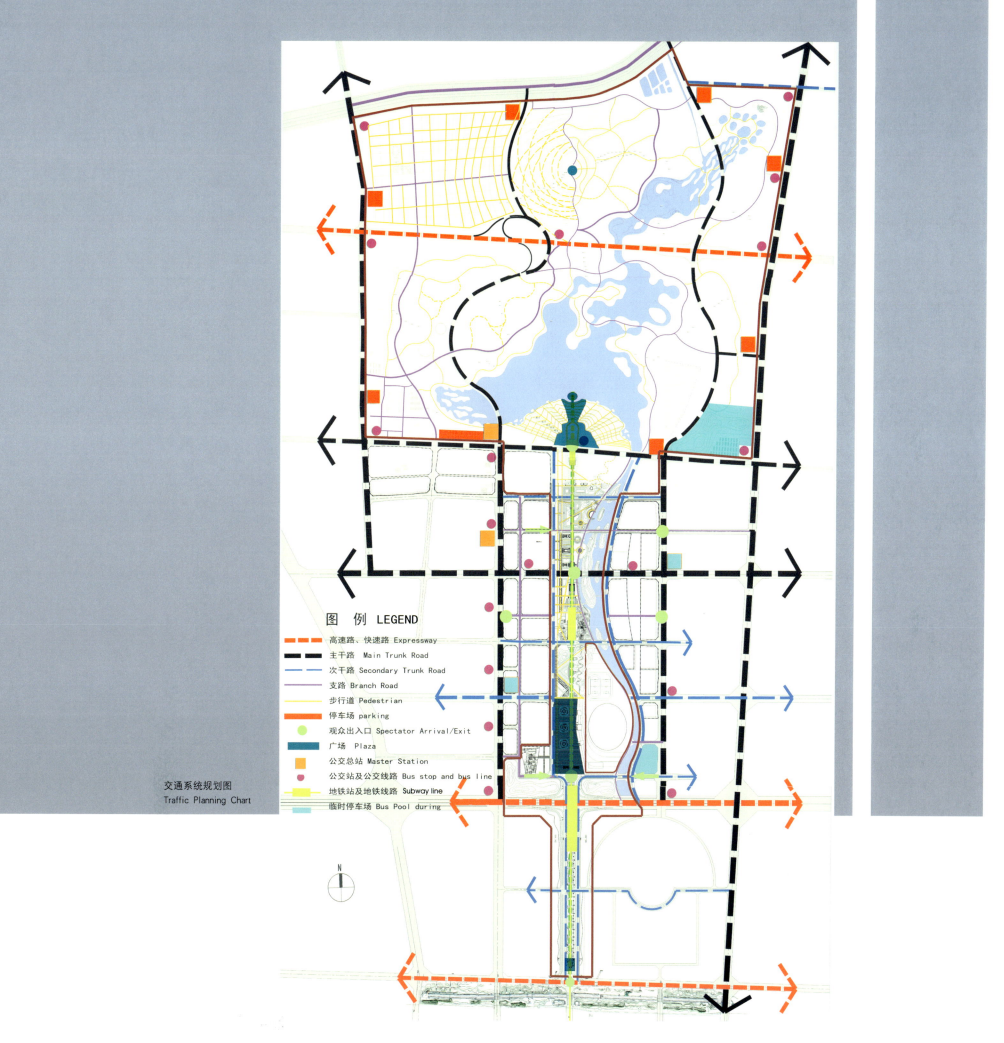

Manature

森林公园竖向规划图
Forest Park Elevation Plan

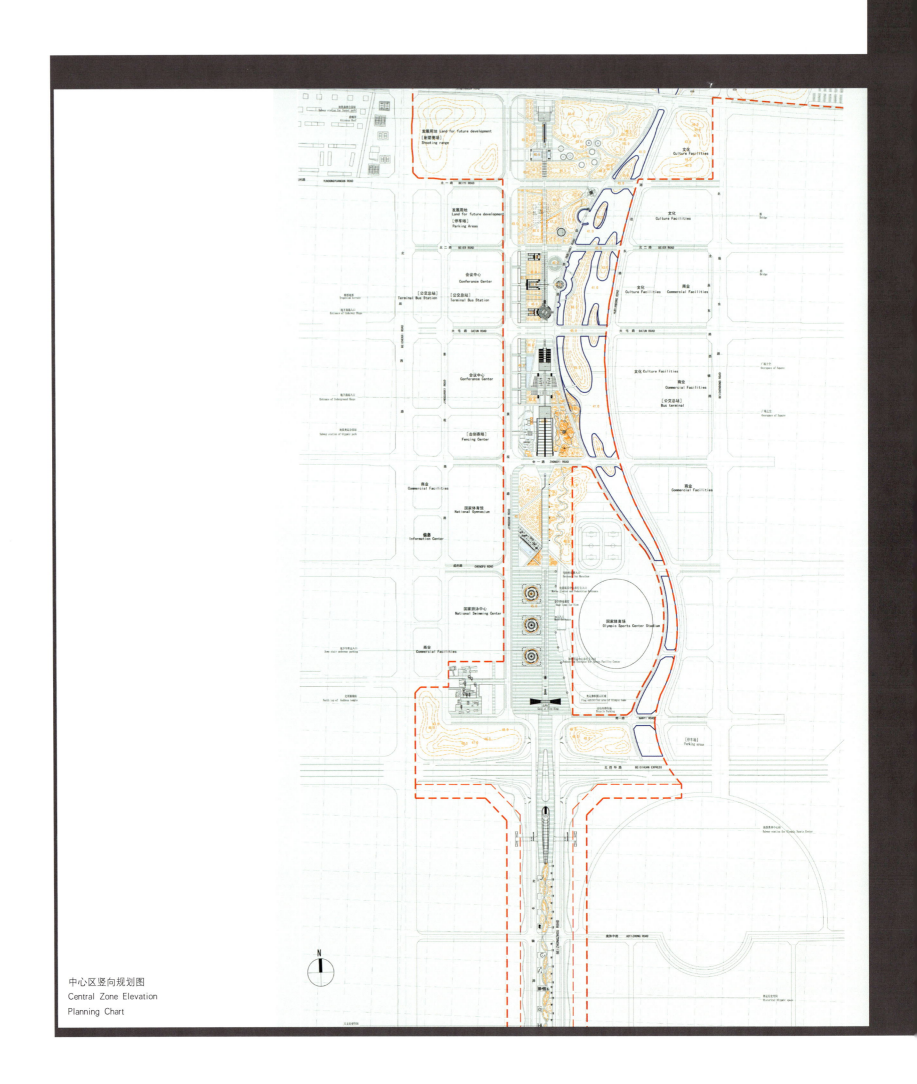

中心区竖向规划图
Central Zone Elevation Planning Chart

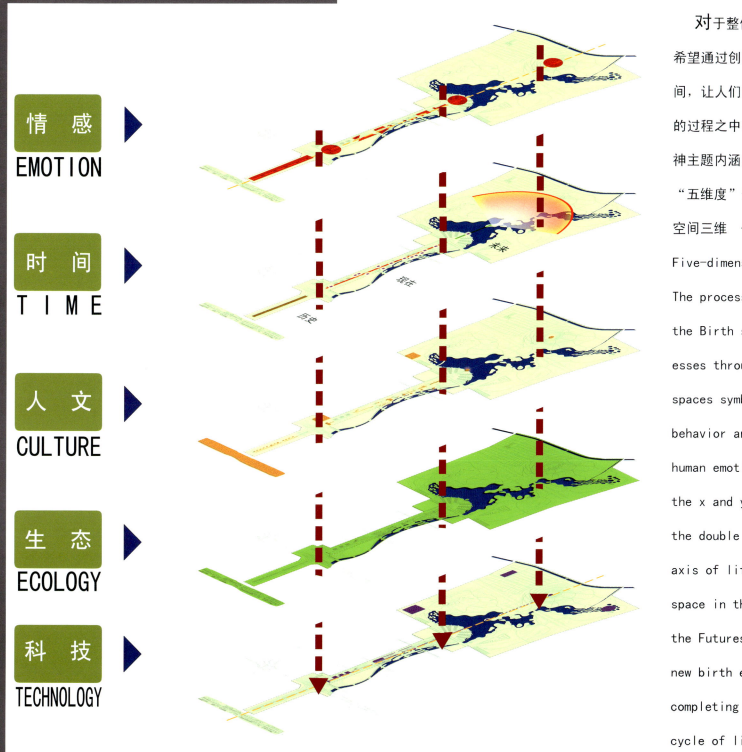

| 情 感 EMOTION |
| 时 间 TIME |
| 人 文 CULTURE |
| 生 态 ECOLOGY |
| 科 技 TECHNOLOGY |

对于整体景观空间，我们希望通过创造有情感主题的空间，让人们在参与和游历空间的过程之中，体会到内在的精神主题内涵。使景观成为一个"五维度"景观：

空间三维 + 时间 + 情感。

Five-dimension Landscapes: The procession begins in the Birth space and progresses through seven paired spaces symbolizing human behavior and complimentary human emotion-coupled as the x and y chromosomes on the double helix, the DNA axis of life, the final space in this sequence is the Futurespace (where a new birth emerges, thus completing and beginning the cycle of life simultaneously)

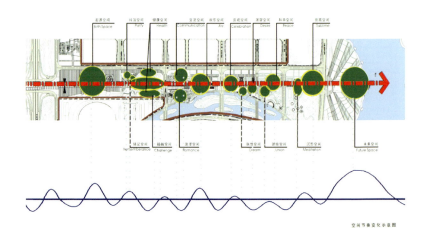

空间节奏变化示意图

CONCEPTION OF THEME SPACES

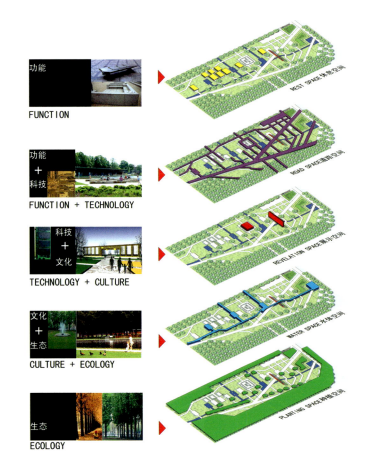

我们希望奥运中心区和森林公园是一个让人们了解奥运历史和精神，并且身体力行感受奥运精神和运动精神的场所空间。所以，我们设计安排了代表人类不断进取和向上的精神的 1+1+7×2个主题精神的情感空间。通过每个主题空间，都能够让人们感受相应的主题精神，并产生不同的情感，在游历物质空间的同时，也是一次情感的空间经历和体验。

We hope the Olympics central zone and forest park give new meaning to history and spirit of the Olympic games. Our motive is to inspire the people with the aspiration of the athletes spirit an understanding of the cumulative experience of visiting the 14 sequential spaces from birth though 7 emotion and 7 behavior space into the future is calculated to up life human spirit beyond the human condition.

The beginning of first kilometer central axis: a space/time line separates yet couples Chinese civilization with Olympics Culture.

文化和文明跨越时空的交流和对话

设计通过对景观元素"时空转换"的处理，将景观的正常时间元素进行错位，产生文化跨越时空对话和交流的效果，并且隐喻中国文化和奥运文化以及世界文明之间潜在的联系。

主题表述 brief of theme spaces

Whole Olympics design project for the center zone and forest park To use the dynamic landscape space of the 'past, present and future' to demonstrate the axis of themes, and arranged cultural with the natural dialogue, which is time and space talk, which announces and expresses Olympic athletics sport to develop the great meaning of human being.

整个奥运森林公园和中心区的设计方案"以象征'历史－现在－未来'的动态景观为主题轴线，安排了一场人文和自然之间跨越时空的对话，通过设计一组表达了运动对于生命成长意义的不同主题序列空间，揭示和表达了奥林匹克体育运动对于人类进步的根本宗旨。"

为了揭示和表达体育运动作为人类进步和发展的动力源泉，追求各种美好情感和事物的过程和作用，我们在中心区主轴线上安排了十六个主题空间：起源、纯洁、铭记……设计创意将人类追求的美好情感的精心归纳总结为纯洁、健康、欢乐、渴望、浪漫、和平、崇高七个主题，反映了人类不同层次情感精神升华的序列。与这七个美好情感主题空间相对应的是七个反映人类为追求这些美好情感和事物而付诸于实践的七种行为：超越、铭记、联想、庆祝、交流、团结、沉思。这7×2主题空间通过起始空间和未来空间联结成一个表达生命发展的主题轴线。

By announcing and expressing the athletic sport as source or impetus to progress by development for mankind, pursuing every kind of perfect emotion with the process or function. On Center Zone axis, we arrange16 theme spaces: birth, purity, remembrance, challenge, health, communication, romance, celebration, joy, peace, dream, enthusiasm, union, sublime, meditation. Future design each space perfect to express the emotion that mankind pursues are seven themes: health, purity, happiness, enthusiasm, romance, peace and sublime, to reflect the preface row that human different level of emotion of the sublimation. Opposite with these seven perfect emotions theme spaces exist another seven theme spaces that is human behaviors in pursuit of these perfect emotions put into practice: challenge, remembrance, dream, celebration, communication, union, meditation. This is 7 by 2 theme spaces. From the birth space to the future space the coupling along the themes axis is an expression life development.

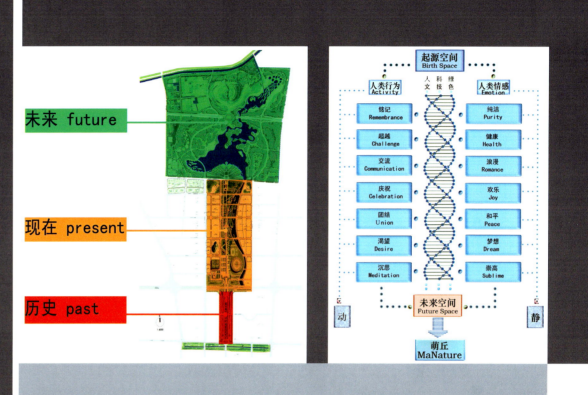

Manature

在五环下团结携手成兄弟。
To union together within five rings of brotherhood.

■ Union Space ／ 团结空间

中国结的造型艺术生动表达了团结的主题。五种色块分别代表了中国五色土以及奥运五大洲家庭。一道镂空人型的弧型墙，邀请人的参与形成手挽手的效果。

The design gives forms according to the art of Chinese knots. Five colors represents the elements: Gold, Wood, Water, Fire, Earth, as well as five races of Olympic rings. A body-shaped engraved wall invites people to participate and understand the meaning of union.

庆祝方式：音乐，喷泉
Celebration Program: Music, Fountain

■ Meditation Space ／ 沉思空间

树庙以及树木封闭的山坳，构成静心沉思的空间。
一排静默水中的卵石，偶尔以上面流出的水滴在水面泛起涟漪，形成水晕。

Tree temples make a nice environment to meditate.
A row of stones unaware of stoneness. Water drops make the ripples and reflections change.

庆祝方式：水纹变化，雾
Celebration Program: Ripple Changes, Fog

未来不可知，
未来可以知。
We are not sure about future,
We can be sure about future.

■ Future Space ／ 未来空间

微地形营造的大地景观，将人引导到湖边。不同的路径暗示"未来不可知"的哲理；沿湖一个大型日晷广场，一半伸入水面，随时间变化，标记下时间，又在强调"未来可以知"的信念。

Soft earth forms leads people to the lakeside. The mesh of pedestrian system gives choices to the walkers, for unpredictable future. Solarium in the square is designed as to record the time change and a landmark to the future theme.

庆祝方式：喷泉，烟火
Celebration Program: Fountain, Firework

■ Dream Space ／ 联想空间

玻璃幕墙创造出光怪陆离的神秘幻觉氛围。人在各个镜子中的不断投影让人产生虚幻的联想。

Mirror Walls give the environment a mysterious motion and seduce people to have dreams.

庆祝方式：光影，雾
Celebration Program: Lighting, Fog

挑战越大，胜利就越值得庆祝。
——托马斯·佩尼
The harder the conflict, the more glorious the triumph.
——Thomas Paine

■ Celebration Space ／ 庆祝空间

用中国传统的鼓、镲、锣、钟及鼓棒等素材构造了一个生动活泼的庆祝空间。

By using Chinese traditional celebration instruments such as drums, chas, bells and sticks as sculpture, the design makes a vivid and interactive celebration space.

庆祝方式：音乐，鼓声，钟声，喷泉
Celebration Program: Music, Drum, Bell, Fountain

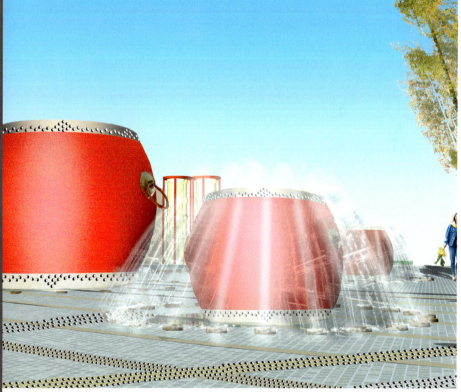

伟大源于联想。
——拉宾维茨
A man's dreams are an index to his greatness.
——Zadok Rabinwitz

生命以为超越而变得精彩.
生命因为超越而更有意义。
Challenges are what make life interesting,
overcoming them is what makes life meaningful.

■ Challenge Space ／ 超越空间

在铺地和墙体上，刻上了体育运动的各种世界记录数据。以一组人跳跃各种障碍的动态雕塑，表达了人类不断超越自己的精神。

With World Records of different games marked on either the wall or pavement, the space design a sequence of motional sculpture to express the challenge spirits.

浪漫始于自爱
To love oneself is the beginning of a lifelong romance

■ Romance Space ／ 浪漫空间

以十二生肖特征创造了十二个互相联系又各自独立的空间，营造相应的环境氛围。公众可以在参与过程中感受浪漫。

We create twelve square to represents twelve scopies with different identities, in plants, facilities. I give people a physical and motional experience to search for romance.

庆祝方式：音乐，鼓声
Celebration Program: Music,Drum

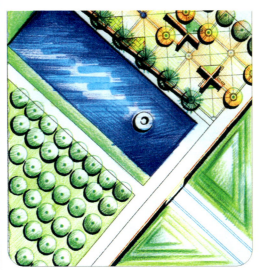

人用一生的磨难回归纯洁
A life cycle returns to purity

■ Purity Space ／ 纯洁空间

一面平静的水面被规则设计的植物环绕。设计以最简洁的设计手法和素材形成一个纯净的空间氛围。

Plants in symmetric form surround a square pool. The design is Minilism and structure ordered in order to create a serenity environment.

铭记让我们重拾记忆，找回人生片刻中永恒的价值。
We don't know the ture value of our moments until they have undergone the test of memory.

■ Remember Space / 铭记空间

错落排列的墙体，为了铭记所有为申办2008和承办2008奉献的无名的有名的英雄。
The Remember Space is created to thank for those heroes, famous or anonymity, worked for Olympic 2008 games.

磨砺弥坚，方始崇高。
——朗费罗
Know how sublime a thing is to suffer and be strong.
—— H. W. Longfe

■ Sublime Space / 崇高空间

崇高空间的地势被局部抬高，形成仰视效果，人通过两道围合的墙体，感受光线从顶部穿透的场景，体会崇高的情愫。
The sublime space is the highest point of the control zone. A closure space between two walls invite people to walk and experience the sublime notions while lights go through the roof.

庆祝方式：落水，水雾
Celebration Program: Falling Water, Fog

欢乐是生命之福，没有欢乐，生命就迷失了方向。
Joy is the only sanction of life; where joy fails, existence remains a mad and lamentable experience.
—— George Santayana

■ Joy Space / 欢乐空间

欢乐空间由两部分组成：一组以中国京剧喜剧脸谱构成的"脸谱"墙，以及模拟笑脸的材质而设计的大地造型。
The joy space is divided into several layers by rotatable wall painted with "Peking Opera Mask". The texture of land cover is also a simulation of Peking Opera Mask.

庆祝方式：转动门，墙
Celebration Program: Rotation of wall to make different masks，Wall

渴望引你至彼岸。
——卡里尔.吉博朗
All that spirits desire, spirits attain.
—— Kahlil Gibran

■ Desire Space / 渴望

螺旋下沉的"黑洞"，以雾、烟、光等虚拟的画面，营造深不可测的渴望探索的场景。
A sunken "Black Hole" with fog, smoke, lighting changing all the times makes a mysterial atmosphere for people desire to searching for infinity.

庆祝方式：音乐，喷泉，雾
Celebration program: Music, Fountain, Fog

运动，开启生命的源头。
Callisthenic, beginning life source

■ Birth Space ／ 起源空间

庆祝方式：灯光，喷水，水柱，火
Celebration Ways: Lighting, Fountain, Water, Fire

和平是惟一的出路。
——A．J．穆斯特
There is no way to peace, peace is the way.
——A. J. Muste

■ Peace Space ／ 和平空间

将子弹头嵌入地下形成的矩阵与树阵对立，体现了和平反战的主题。树从两堵墙的夹缝中长出，下面是断枪支起的支架，表现了和平产生新生命的蕴涵深意。

A pattern of 'bullets' on one side, a forest Matrix on the other side; trees grow up between two walls, from a abandoned guns as bracket. It strongly emphasized the theme of peace and life.

庆祝方式：喷泉，音乐
Celebration Program: Fountain, Music

交流是浪漫之源
交流是灵感之泉
Communication Arouses Romance
Communication Creates Inspiration

■ Communication Space / 交流空间

若干汇聚场地，通过水的流域组织形成网络结构，既构成了丰富的景观，又寓意信息交流产生生命的主题。整个场地设计，无论材质，质感和动势表现，借鉴高科技信息材料如电脑芯片的形态、质感，营造了高科技相关的教育空间。

The meting places are connected by pathway and water corridor. A network of streamlines and pathways makes the interlinked nodes interesting space. The design simulates the form, texture and structure of some information technology materials, such as Ohips. It also provides education of high technology knowledge and philosophy to the public.

庆祝方式：喷泉，流水速度
Celebration Program: Fountain, Water Flowing Speed

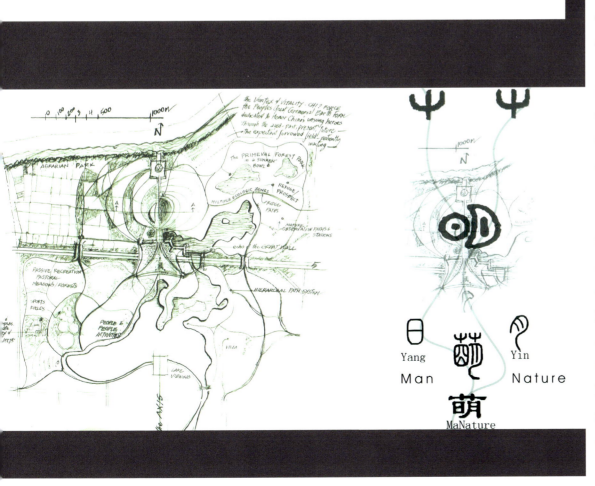

在主题空间的表达方式上，我们借用了强烈显示了中国文化背景的元素来表达设计主旨，烘托环境气氛，并以此达到诠释主题的目的。对于主题空间意象的选择，我们则力图将中国文化的内容变成一个全球性的（universal）的语言。
In the theme spaces abstraction of Chinese culture and spirit are given form to convey universal meaning to the people.

由中国文字"萌"（日月（阴阳）衍生万物）演绎而来的设计表达，又是英文man和nature合成的MaNature，表达"天人合一"的含义。对于设计概念的文化阐释，既是中国的，又是世界的。
Man/nature Manature: two Chinese character MEN QUI the word MEN in Chinese is sun/moon/grass combine as MEN which suggest coining the English manature ancient Chinese philosophy directs us to express culture/nature in our design.

中国文化元素
Chinese Culture Element

墙和门的序列
A Preface Sequence of Wall with Gate

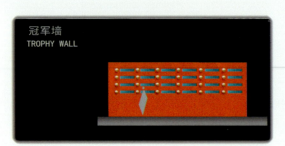

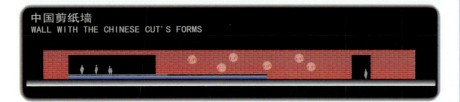

1： 作为记录奥运历史的载体
   历史记录
   动态记录 2008 届奥运
Garriers the historical record of the Olympic games History record
The development records 2008 Olympics

2： 作为记录文明的载体
Each day, the Olympic medallist are memorialized in the great wall

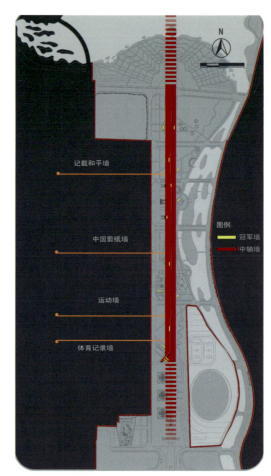

墙的序列
A Preface Sequence of Wall

"九重门"使中轴线景观层次更加丰富，不同情感主题的窄界定更加明确。
The paradise 9 consecufive gate define the space sequence along the axis as shown．

3：作为分隔空间的景观
A design device to create a sequence of landscape spaces

4：作为连接延伸景观空间物体（建筑－园林，园林－园林）
An object to link or extend the landscape space(building-park,park-park)

5：作为造景的元素(透景／漏景／框景／对景／视觉焦点)
An element to build landscape (active scenery/borrowedscenery/enflamed scenery/appositive scenery/sense of vision focus)

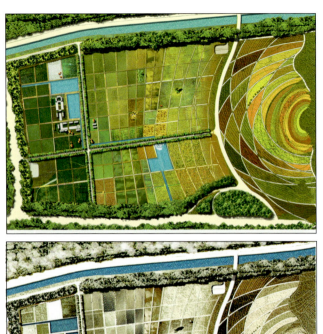

四季变化的风景
Seasons

A06 Beijing Park Society

北京风景园林协会

Beijing China Research Center of Landscape Architectural Design and Planning
Beijing Topsense Landscape & Design Co.,Ltd.
Tianxia Original Color Design Ltd.
The Landscape Architecture School of Beijing Forestry University
Mark VanZeumcren,P.Eng.
北京中国风景园林规划设计研究中心
北京创新景观园林设计有限责任公司
北京天下原色艺术设计有限责任公司
北京林业大学园林学院
加拿大马克·凡泽梅伦公司

奥林匹克公园森林公园及中心区景观规划设计方案

OLYMPIC GREEN LANDSCAPE DESIGN FOREST PARK AND CENTRAL ZONE

奥林匹克森林公园及中心区全园鸟瞰
Birds-eye View of Forest Park and Central Zone

奥林匹克公园
森林公园和中心区景观规划方案概述

一、体现古都传承——融入自然的中轴

1) 北京中轴线作为城市文脉精髓得到尊重。在中心区景观序列沿中轴线展开，两侧林带也由入口区行列式过渡到公园入口区的微地形自然式，在森林公园通过喷泉与灯光，将中轴融入自然山水之中。

2) 中轴线以平直的游步道贯穿整个中心区，西侧种植银杏林带形成气势，林下绿地抬起，其边缘设计为曲线形五级台阶，与游步道形成对比；东侧水岛绿地采用自然式种植，呼应中心区空间开合变化。

3) 游步道铺装采用御道符号形式，强调中轴线"龙脉"的象征性。

二、以活跃的绿色连接二大功能区——飘进城市的绿洲

中心区与森林公园通过三种活跃的景观形式连为一体：

1) 东侧水系中以自然式植物群落为主的绿岛系列。

2) 公园入口广场的丘陵绿地，形成森林公园山脉在中心区的延续。

3) 中心区广场之间安置众多自然式种植岛，并与规整的林带形成对比。

三、表现奥运精神——激情同享的奥运

通过中心区五个节点广场以及奥运雕塑园、奥运纪念园，体现奥运精神与人文关怀，特别注重人性化空间与设施的设计，曲线形五级台阶贯穿整个中心区，增加近人尺度，创造"广泛参与"的奥运氛围；同时台阶五色代表奥运五环与中国五色土，寓意东西文化的交流。

四、建立生态绿洲——天人和谐的森林

森林公园将建成一个绿色环保公园。人与自然协调共生是森林公园设计的终极目标。

1) 森林公园基本风貌为以人工模拟的森林景观，五环路北为山林，五环路南为森林边的湖泊、沼泽景观。

2) 湿地与双重水源过滤系统。

标书名称	表达形式	表达的内涵	广场名称	设计景观	说明
入口广场	奥运会徽	舞动的北京	"舞"（篆刻形式）	1.中国印 2.喷泉广场	规整的林带形成强烈导向
体育广场	奥运圣火	更快、更高、更强	"火"（篆刻形式）	1.五组火炬喷泉 2.林下运动场地、器械 3.引发运动冲动的空间	1.留出充足人流集散场地 2.北顶庙前辟为民族体育园
交通广场	奥运会旗	友谊、团结、和平 相聚在五环旗下	"爱"（篆刻形式）	1.旗杆阵 2.易于交往的各类空间 3.回音墙、留言墙	1.交通—交流—交谈
文化休闲广场	奥运会歌	美与尊严 冠军在你我之间	"美"（篆刻形式）	1.本届获奖者纪念小品 2.领奖台式休闲家具	1.水岛相连，临水广场 2.街头表演：绘画、演奏
公园入口广场	奥运吉祥物	人与万物共生	"绿"（篆刻形式）	1.草地看台 2.花卉专类园	1.绿色丘陵环抱 2.婚礼

效果图 Effect Drawing

The Olympic Park

The Summary of the Plan of the Design of the Forest Park and the Scene in the Central Area

I. Embodying the heritage of the ancient capital—the central axis being mixed into nature

1) The central axis of Beijing gains respect as the soul of the city's cultural heritage. The sequence of the scenes in the central area is expanded along the central axis. The forest belt of the two sides also changes from the determinant pattern in the entrance area into the natural form of tiny landform in the entrance area of the park. The central axis is blended into the natural mountains and waters through the use of fountains and lamplight in the Forest Park.

2) The central axis runs through the whole central area in the form of even and straight footpath. The gingko belt, planted in the west side, becomes overwhelming. The greenbelt under the forest is elevated. The margin of it is designed as curving five-stage steps, forming a contrast with the footpath, increasing the scale of closing to people. The greenbelt of the island to the east side adopts the natural planting pattern, echoing with the spatial changes of folding and unfolding in the central area.

3) The pavement of the footpath adopts the pattern used by imperial paths, emphasizing the central axis' symbolic heritage of dragon.

II. Linking the major two functional areas through the use of the energetic greenness — the oasis floating into the city

The central area and the Forest Park are combined into a single system through the use of three energetic scenic patterns.

1) The series of green islands primarily occupied by natural vegetation community in the water system of the east side.

2) The hilly greenbelt in the square of the park's entrance forms the continuance of the Forest Park's mountain chain in the central area.

3) Numerous natural pattern planting islands are arranged between the squares in the central area, forming a contrast with the neat forest belt.

III. Embodying the Olympic spirits — the shared passion for Olympic Games

The Olympic spirits and human thought are embodied through the five node squares in the central areas, the Olympic Sculpture Park, and the Olympic Commemoration Park. The design of human space and facilities is given primary attention. The curving five-stage steps run through the whole central area, increasing the scale of closing to people, creating an Olympic atmosphere of extensive participation. At the same time, the five colors of the steps represent the Olympic five rings and the Chinese five-color earth, implying the cultural communication between the east and the west.

IV. Establishing zoological oasis—the forests where nature and human are in harmony

Forest Park will be built as a green environment protected park. The ultimate goal of the design of the Forest Park is to contribute to the harmony between human and nature.

1) The basic scene of Forest Park is the artificial forest scene. To the north side of the Fifth Ring Road are forests. To the south side of the Fifth Ring Road are the scenes of lakes and marshes.

2) The purifying system of everglade and double water sources.

The everglade adopts the form of terrace, declining layer upon layer, intersecting and blending with mountains and

Name of the mark	Form of representation	The represented connotations	Name of the squares	The designed sights	Explanations
Square of the entrance	The Olympic emblem	Dancing Beijing	Dancing (seal cutting)	1. Chinese seal 2. The square of fountains	The neat forest belt forms strong guidance
Sports Square	Sacred fire for Olympic use	Quicker, higher, Stronger	Fire (seal cutting)	1. Five series of torch-shape fountains 2. Sports field and facilities under forests 3. The space causing people have the impulse to have exercises	1. Leaving enough fields for the assembly and dispersal of the stream of people 2. The national sports park is established in the front of the North Peak Temple
Traffic Square	The Olympic flag	Friendship, solidarity, peace Assembling under the five-ring flag	Love (seal cutting)	1. An array of masts 2. Various spaces facilitating communication 3. Echo wall, wall for leaving words	Transportation - Communication - Conversation
Cultural Square for relaxation	The Olympic song	Beauty and Respect The champion is between you and me	Beauty (seal cutting)	1. Souvenirs for the champions 2. Informal furniture resembling the platform for awarding	1. Water and islands are connected, the square close to water 2. Performances in the streets: painting and musical performances
Square of the entrance of the park	The Olympic mascot	Human and animals are living together	Greenness (scal cutting)	1. The stand built on the lawn 2. Specialized garden for flowers	1. Encircled by green hilly land 2. Wedding

* 湿地采用梯田形式，层层跌落并与山林穿插交融。三种水源(中水、雨水、河湖水)引入湿地过滤处理后注入森林公园湖泊。湿地分为大小两部分，大面积湿地位于山脉环抱之中，小面积湿地展现湖面。
　　* 双重过滤系统：中水过滤系统与湖水过滤自循环系统。
　　大面积湿地用来过滤中水，并形成峡谷与缓滩景观连接大湖。小面积湿地用来过滤经水泵提升的湖水，形成过滤自循环系统。
　　* 湿地上游建立温室，开展生态教育与湿地生物展示。
　3）仰山大沟改造为动植物栖息谷地，并与中心区水岛连通，成为城市生态廊道。
　4）植物群落分为山阳、山阴、水岸、湿地 4 类 12 个组合单元，确保植物的多样性，便于操作施工。
　5）山林中以"林窗"形式布置游人活动空间，每个空间结合原有村庄肌理、街坊基址建立游客服务中心。

五、延续地域文脉
　1）将众多的历史遗存融入景观之中，使公园成为一个唤起联想、感染亲情的空间，形成一个继往开来的自然人文环境。
　　* 龙王堂：先民对水源地——仰山洼的尊重、敬畏与祈盼。
　　* 北顶娘娘庙：当地先民盛大的聚会，祈盼青春女神的祝福。
　　……元大都遗址、墓碑群、16 处古老村名、村庄肌理、湖水、湿地梯田。
　2）建立"印记——历史遗存走廊"，始自"舞"之广场，将各处历史遗存串连成环。
　　"历史遗存走廊"设计包括四要素：历史遗迹、石板游步道、松科林阴廊、解说小品。

六、运用中国园林的布局手法
　1）借鉴颐和园挖湖堆山手法安排公园的山水骨架、借景西北远山。
　2）继承意境的创作方式，为景区题名标识。
　3）运用"因借"地势环境手法，安排森林公园中的奥运纪念园、景观建筑。
　4）使用堆山手法进行湖岸的局部处理。

七、本方案所表达的理念：

融入自然的中轴
飘进城市的绿洲
激情同享的奥运
天人和谐的森林
地域文脉的延续
中国园林的布局

forests. Three water sources (normal water, rain, rivers and lakes) are infused into the lakes in Forest Park after being purified in the everglade. The everglade can be separated into two parts, a big one and a small one. The big one lies in the surrounding of the mountains, while the small one shows the surface of the lake.

The double purifying systems: the purifying system of normal water and the self-circulated purifying system of lakes.

The big one is used to purify normal water, forming the scene of canyon and beach that combine big lakes. The small one is use to purify the lake water got by using water pump, forming self-circulating purifying system.

Establishing greenhouse in the upper reaches of the everglade, developing zoological education and the exhibition of living things living in everglade.

3) The Yangshan Channel is transformed to a valley inhabited by animals and vegetation. Connected with the islands in the central area, it will become the city's zoological corridor.

4) The vegetation community is divided into yang, yin, bank, everglade, altogether four kinds and twelve units, insuring the diversity of the plants, convenient for manipulation and construction.

5) In forests, the tourists' activity area is arranged in the format of trees and windows. Combined with the former village patterns and locations of streets, every area establishes service center for tourists.

V. Continuing the regional heritages

1) Taking numerous historical heritages into the scene, the park becomes a place that can evoke imagination and communicate benignity, forming a natural human environment carrying forward the cause and forging ahead into the future.

The Hall of the King of dragon: the respect, awe, and pray of the ancient people towards the water source—Yangshan Channel.

The North Peak Temple of the Goddess of Fertility: the grand assembly of the local ancient people, praying the bless of the Youth Goddess.

……the site of Yuan Dynasty, the tombstone groups, sixteen names of ancient villages, format of the villages, lake, the terrace of the everglade.

2) Establishing marking—the corridor of historical heritages. Originating from the Square of Dancing, the historical heritages of various places will be linked.

The design of the corridor of historical heritages includes the following elements: historical heritages, slate footpath, corridor with pine forests, explanations.

VI. Adopting the arranging methods of Chinese gardens

1) Arranging the framework of the landscape in the park by using the Summer Palace's method of digging lake and piling mountains for reference, borrowing the scene from the far northwest mountains.

2) Making superscriptions for the scenic areas by inheriting the creation pattern using artistic conception.

3) Arranging the Olympic Commemoration Park and the scenic buildings by using the methods of borrowing hypsography and environment.

4) Arranging the partial treatment of the lakeshore by using the method of piling the mountains.

VII. The conceptions represented by this plan

the central axis being mixed into nature

the green oasis floating into the city

the shared passion for Olympic Games

the forest where nature and human are in harmony

the continuance of the regional heritage

the arrangement of the Chinese gardens

奥林匹克森林公园及中心区全园平面图 [赛时]
Plan of Forest Park and Central Zone (during Olympics)

天人共荣的森林
forests where nature and human are in harmony

建立生态绿洲
Establishing zoological oasisthe forests

山林中以"林窗"形式布置游人活动空间
In forests, the tourists' activity area is arranged in the format of trees and windows.

湿地与双重水源过滤系统
The purifying system of everglade and double water sources

森林公园基本风貌为以人工模拟的森林景观
The basic scene of Forest Park is the artificial forest scene.

融入自然的中轴
The centrad axis melted into nature

体现古都传承融入自然的中轴
Embodying the heritage of the ancient capitalthe central axis being mixed into nature

在中心区景观序列沿中轴线展开。两侧林带也由入口区行列式过渡到公园入口的微地形自然式,在森林公园通过喷泉与灯光,将中轴融入自然山水之中。
The sequence of the scenes in the central area is expanded along the central axis. The forest belt of the two sides also changes from the determinant pattern in the entrance area into the natural form of tiny landform in the entrance areaof the park. The central axis is blended into the natural mountains and waters through the use of fountains and lamplight in the Forest Park.

地域文脉的延续
Continuing the regional heritages

建立"印记历史遗存走廊",将各处历史遗存串连成环。
Establishing markingthe corridor of historical heritages. the historical heritages of various places will be linked.

龙王堂现状 Dragon King Hall

改造后效果图 Effect Drawing

娘娘庙现状 Beijing Nianguiang Temple

改造后效果图 Effect Drawing

飘进城市的绿洲
Greenness the oasis floating into the city

以三种绿岛形式连接中心区与森林公园
Linking the major two functional areas through the use of the energetic greennessthe oasis floating into the city

激情同享的奥运
Embodying the Olympic spiritsthe shared passion for Olympic Games

通过中心区五个节点广场体现奥运精神与人文关怀
The Olympic spirits and human thought areembodied through the five node squares in the central areas

入口广场-奥运会徽
-舞动的北京-舞之广场
Square of the entrance
-The Olympic emblem
-Dancing Beijing
-Dance Plaza

体育广场-奥运圣火
-更快、更高、更强-火之广场
Sports Square
-Sacred fire for Olympic use
-Quicker, higher, Stronger
-Fire Plaza

交通广场-奥运会旗
-友谊、团结、和平-爱之广场
Traffic Square
-The Olympic flag
-Friendship, solidarity, peace
-Love Plaza

文化休闲广场-奥运会歌
-美与尊严-美之广场
Cultural Square for relaxation
-The Olympic song
-Beauty and Respect
-Beauty Plaza

公园入口广场-奥运吉祥物
-人与万物共生-绿之广场
Square of the entrance of the park
-The Olympic mascot
-Human and animals are living together
-Green Plaza

设计理念
Design Concept

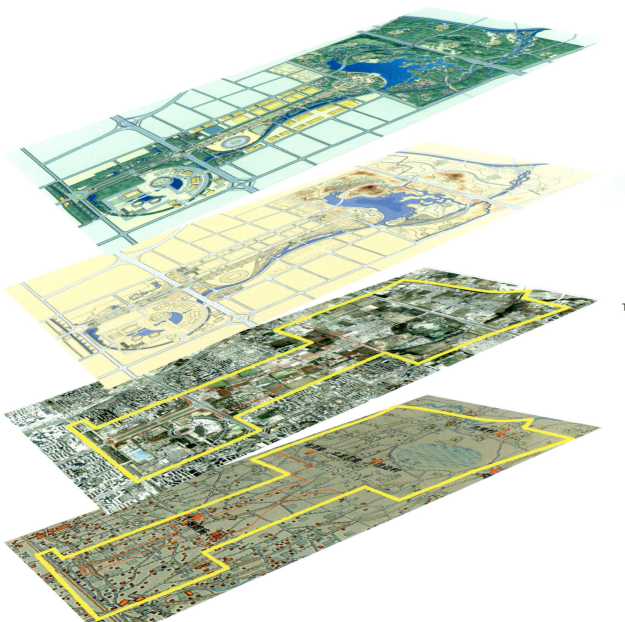

规划总平面 2008年
General Layout

山水骨架 2004年
The Framework of Mountains and Rivers

现状航拍图 2002年
The Aerial Photograph of Current Situation

历史渊源 1930年
Historical Origins

Design Concept

设计理念

1930年普意雅1：2500测绘图上显示
的仰山洼湖水、稻田与龙王堂
Yangshawa Lake, rice-field, and Dragon King
Temple shown in 1930 Puya map (1:2500)

2008奥林匹克公园水系-龙海
2008 Olplc Park water system-Dragon

人工湿地使洼里地区成为
重要的贡米产地。
The area inside Yangshawa is the
chief growing place of rice of
tribute due to the man-made wetland

模拟梯田的人工湿地。
The man-made wetland that imitates the terrace.

建于明代的龙王堂显示先民
对湖水洼田的尊重与敬畏。
Dragon King Temple built in the
Ming Dynasty shows the respect,
awe, and veneration of the
ancient people to lake and
low land.

建设森林公园是对生态环境尊重，
是新一轮的回归。
龙王堂是历史之源。
To construct the forest park is to respect the ecological
environment as well as a new round of return. Dragon
King Hall is the source of history.

当地先民隆重聚会的所在
The place the local ancient people gathered

奥运是新世纪国际盛会。
The Olympic Game is the international
gathering of the new century.

古人对水草丰美环境的认证与眷恋。
The acknowledgement and affection of the ancient
people for the fertile land

"印迹走廊"说不尽的故事
"The corridor of historical marks" and endless stories.

人的痕迹
Trace of the human race

山林中以"林窗"形式布置游人活动空间，
每个空间结合原有村庄肌理，街坊基址建
立游客服务中心
The active space for visitor is arranged by the type
of "forest window" in mountain forest,Each space
combine with the original village's texture. The
visitor information center is build on the base site
of courtyard

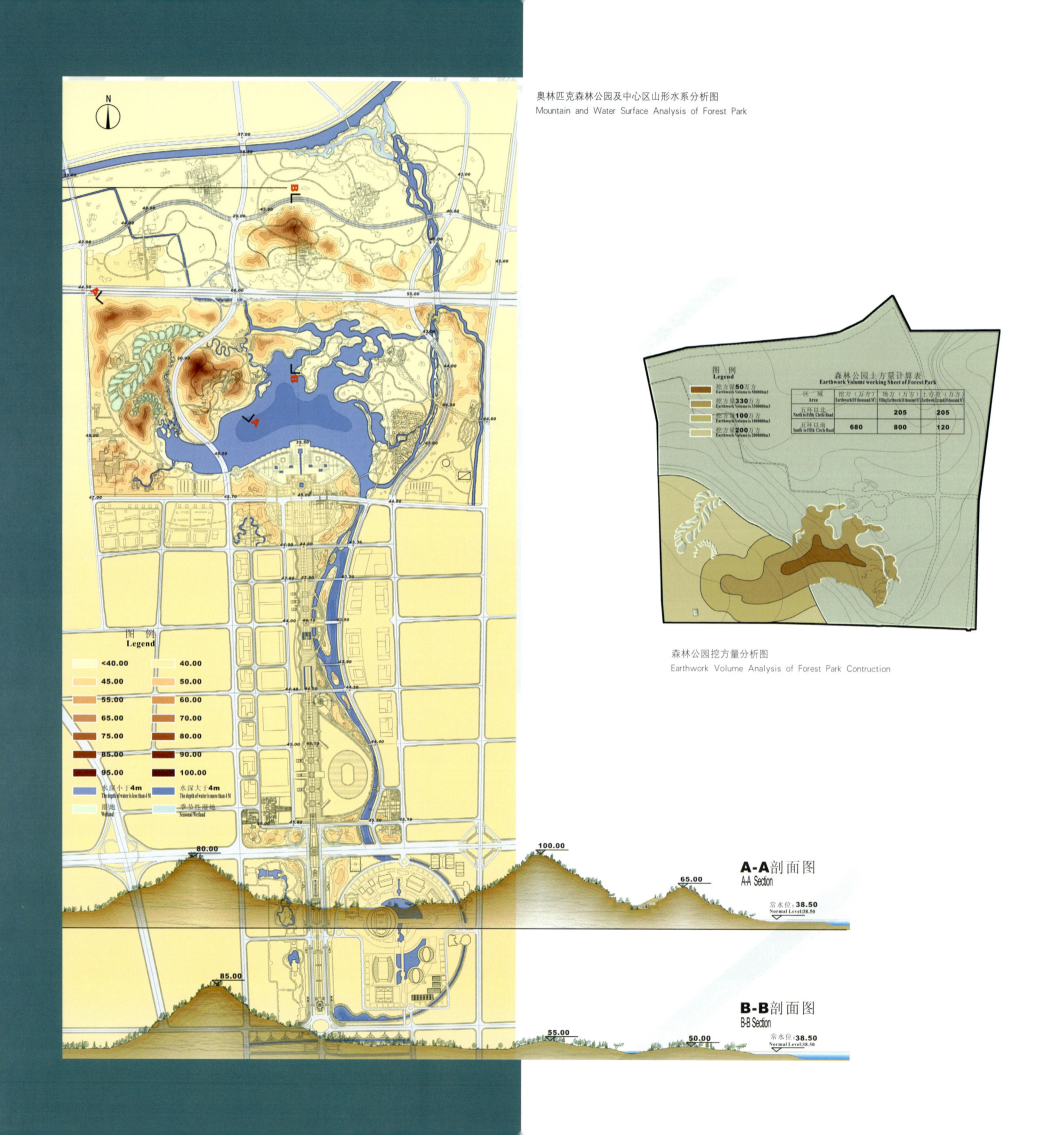

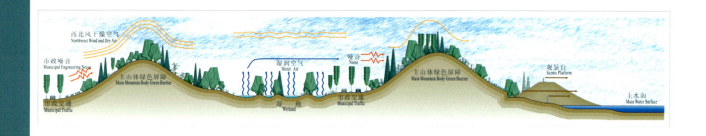

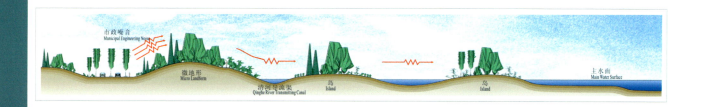

森林公园地形水系生态效果分析图
Mountain and Water Surface Sectional Relationship and Impact Analysis of ForestPark

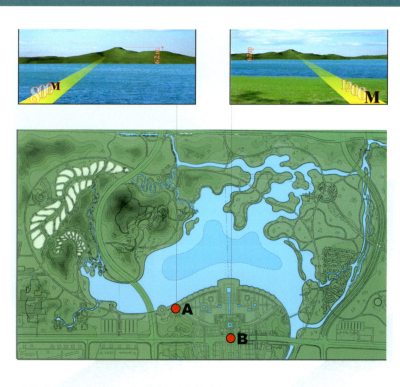

森林公园湖边主峰与颐和园万寿山视觉分析对比
Visual Analysis Comparison between Lakeside Main Peak and Summer Palace Garden, Wanshou Mountain

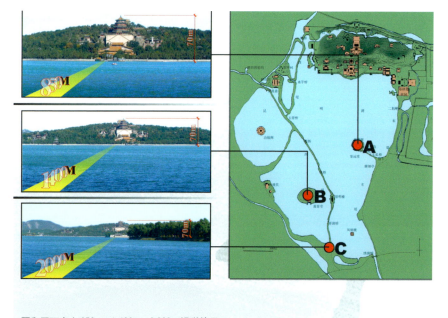

颐和园万寿山850m、1 400m、2 000m视觉效果
Visual Impact of Wanshou Mountain at Distance of 850m, 1 400m and 2 000m, Summer Palace

森林公园湖边主峰800m、1 200m视觉效果
Visual Impact of Lakeside Main Mountain at Distance of 800m, 1 200m, Summer Palace

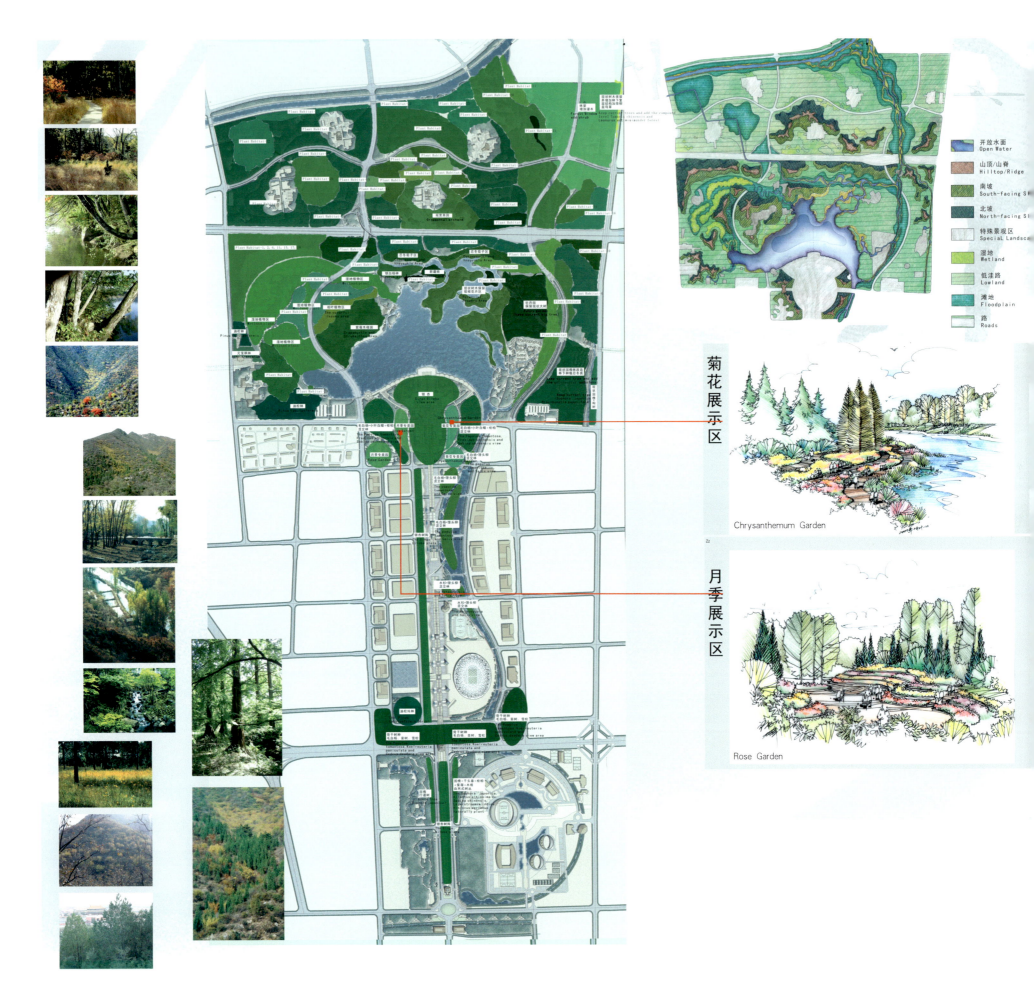

菊花展示区

月季展示区

Chrysanthemum Garden

Rose Garden

奥林匹克森林公园及中心区种植规划方案
Planting Plan of Forest Park and Central Zone

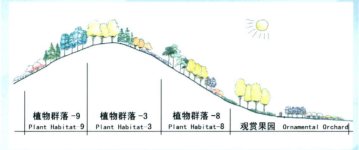

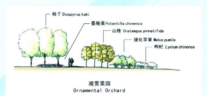
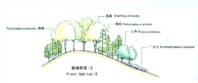

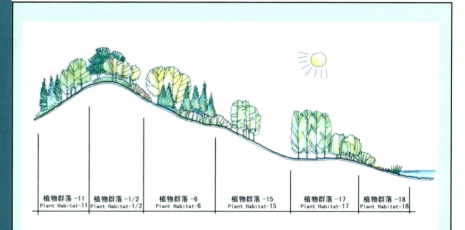
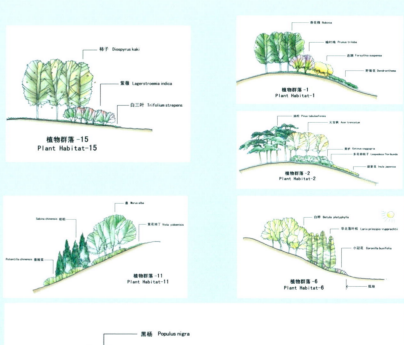
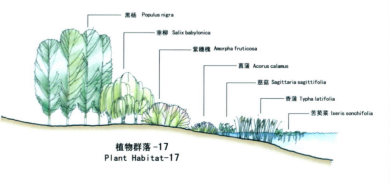

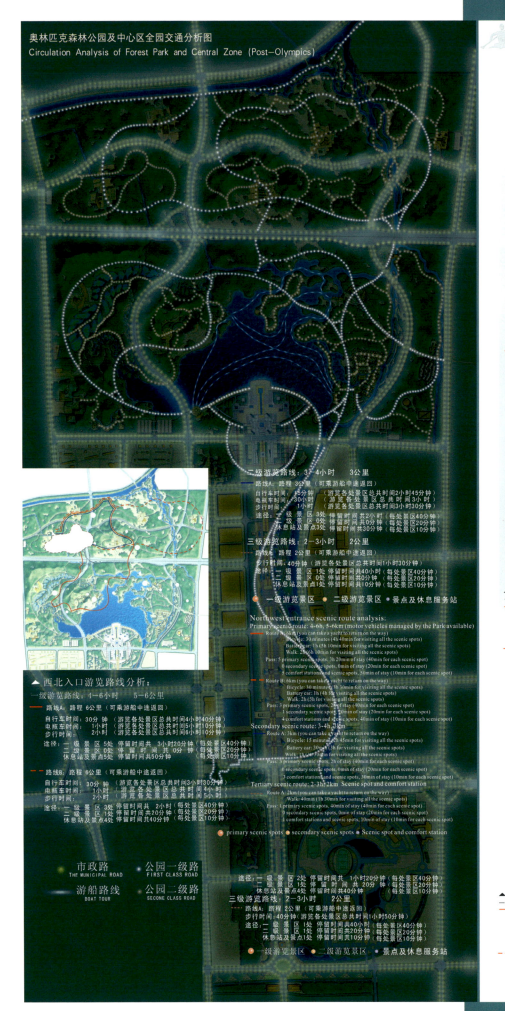

◀ 西南入口游览路线分析：
一级游览路线：4—5小时　5—6公里（可行驶公园管理机动车）
　　路线A：路程 6公里（可乘游船中途返回）
　　　　自行车时间：30分 钟 （游览各处景区总共时间3小时50分钟）
　　　　电瓶车时间：1小时 （游览各处景区总共时间4小时20分钟）
　　　　步行时间：2小时 （游览各处景区总共时间5小时20分钟）
　　　　途径：一级景区3处 停留时间共 2小时（每处景区40分钟）
　　　　　　　二级景区2处 停留时间40分钟（每处景区20分钟）
　　　　　　　休息站及景点4处 停留时间40分钟（每处景区10分钟）
　　路线B：路程 6公里（可乘游船中途返回）
　　　　自行车时间：30分 钟 （游览各处景区总共时间3小时10分钟）
　　　　电瓶车时间：1小时 （游览各处景区总共时间3小时40分钟）
　　　　步行时间：2小时 （游览各处景区总共时间4小时40分钟）
　　　　途径：一级景区4处 停留时间共 1小时40分钟（每处景区40分钟）
　　　　　　　二级景区1处 停留时间共 20分钟（每处景区20分钟）
　　　　　　　休息站及景点4处 停留时间共40分钟 （每处景区10分钟）
二级游览路线：3—4小时　3公里
　　路线A：路程 3公里（可乘游船中途返回）
　　　　自行车时间：15分钟 （游览各处景区总共时间2小时35分钟）
　　　　电瓶车时间：30小时 （游览各处景区总共时间2小时50分钟）
　　　　步行时间：1小时 （游览各处景区总共时间3小时20分钟）

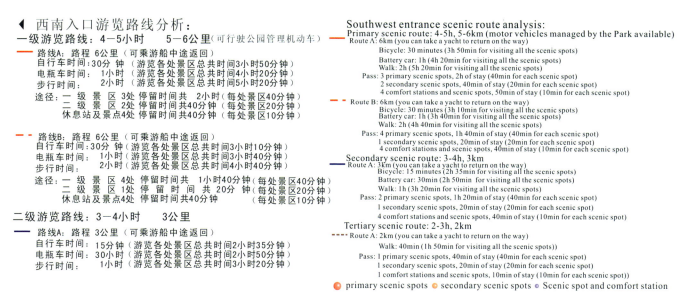

Southwest entrance scenic route analysis:
Primary scenic route: 4-5h, 5-6km (motor vehicles managed by the Park available)
— Route A: 6km (you can take a yacht to return on the way)
　　Bicycle: 30 minutes (3h 50min for visiting all the scenic spots)
　　Battery car: 1h (4h 20min for visiting all the scenic spots)
　　Walk: 2h (5h 20min for visiting all the scenic spots)
　　Pass: 3 primary scenic spots, 2h of stay (40min for each scenic spot)
　　　　2 secondary scenic spots, 40min of stay (20min for each scenic spot)
　　　　4 comfort stations and scenic spots, 50min of stay (10min for each scenic spot)
--- Route B: 6km (you can take a yacht to return on the way)
　　Bicycle: 30 minutes (3h 10min for visiting all the scenic spots)
　　Battery car: 1h (3h 40min for visiting all the scenic spots)
　　Walk: 2h (4h 40min for visiting all the scenic spots)
　　Pass: 4 primary scenic spots, 1h 40min of stay (40min for each scenic spot)
　　　　1 secondary scenic spots, 20min of stay (20min for each scenic spot)
　　　　4 comfort stations and scenic spots, 40min of stay (10min for each scenic spot)
Secondary scenic route: 3-4h, 3km
— Route A: 3km (you can take a yacht to return on the way)
　　Bicycle: 15 minutes (2h 35min for visiting all the scenic spots)
　　Battery car: 30min (2h 50min for visiting all the scenic spots)
　　Walk: 1h (3h 20min for visiting all the scenic spots)
　　Pass: 2 primary scenic spots, 1h 20min of stay (40min for each scenic spot)
　　　　1 secondary scenic spots, 20min of stay (20min for each scenic spot)
　　　　4 comfort stations and scenic spots, 40min of stay (10min for each scenic spot)
Tertiary scenic route: 2-3h, 2km
--- Route A: 2km (you can take a yacht to return on the way)
　　Walk: 40min (1h 50min for visiting all the scenic spots)
　　Pass: 1 primary scenic spots, 40min of stay (40min for each scenic spot)
　　　　1 secondary scenic spots, 20min of stay (20min for each scenic spot)
　　　　1 comfort stations and scenic spots, 10min of stay (10min for each scenic spot)

● primary scenic spots　● secondary scenic spots　● Scenic spot and comfort station

Circulation

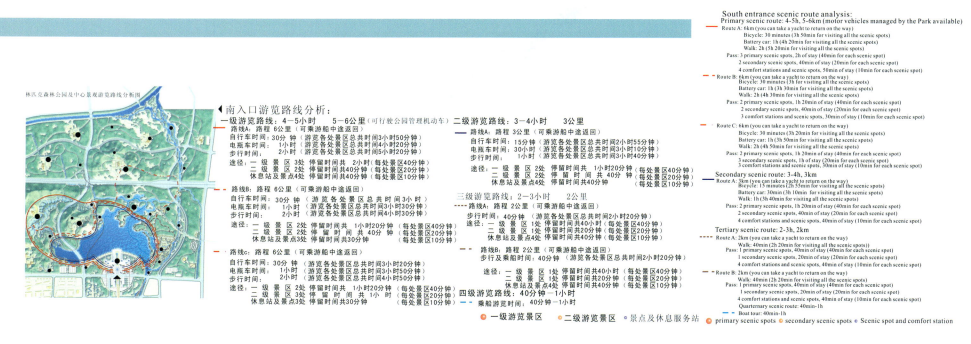

◀ 南入口游览路线分析：
一级游览路线：4—5小时　5—6公里（可行驶公园管理机动车）
　路线A：路程 6公里（可乘游船中途返回）
　　自行车时间：30分钟 （游览各处景区总共时间3小时50分钟）
　　电瓶车时间：1小时 （游览各处景区总共时间4小时20分钟）
　　步行时间：2小时 （游览各处景区总共时间5小时20分钟）
　　途径：一级景区3处 停留时间共 2小时（每处景区40分钟）
　　　　　二级景区2处 停留时间1小时20分钟（每处景区40分钟）
　　　　　休息站及景点4处 停留时间40分钟（每处景区10分钟）
　路线B：路程 6公里（可乘游船中途返回）
　　自行车时间：30分钟 （游览各处景区总共时间3小时）
　　电瓶车时间：1小时 （游览各处景区总共时间3小时30分钟）
　　步行时间：2小时 （游览各处景区总共时间4小时30分钟）
　　途径：一级景区2处 停留时间 1小时20分钟（每处景区40分钟）
　　　　　二级景区2处 停留时间 40分钟（每处景区20分钟）
　　　　　休息站及景点3处 停留时间共30分钟（每处景区10分钟）
　路线C：路程 6公里（可乘游船中途返回）
　　自行车时间：30分钟 （游览各处景区总共时间3小时20分钟）
　　电瓶车时间：1小时 （游览各处景区总共时间3小时50分钟）
　　步行时间：2小时 （游览各处景区总共时间4小时50分钟）
　　途径：一级景区2处 停留时间共 1小时20分钟（每处景区40分钟）
　　　　　二级景区3处 停留时间共 1小时（每处景区20分钟）
　　　　　休息站及景点3处 停留时间共30分钟（每处景区10分钟）

二级游览路线：3—4小时　3公里
　路线A：路程 3公里（可乘游船中途返回）
　　自行车时间：15分钟 （游览各处景区总共时间2小时55分钟）
　　电瓶车时间：30分钟 （游览各处景区总共时间3小时10分钟）
　　步行时间：1小时 （游览各处景区总共时间3小时40分钟）
　　途径：一级景区2处 停留时间共 1小时20分钟（每处景区40分钟）
　　　　　二级景区2处 停留时间共 40分钟（每处景区20分钟）
　　　　　休息站及景点4处 停留时间40分钟（每处景区10分钟）

三级游览路线：2—3小时　2公里
　路线A：路程 2公里（可乘游船中途返回）
　　步行时间：40分钟 （游览各处景区总共时间2小时20分钟）
　　途径：一级景区1处 停留时间共40分钟（每处景区40分钟）
　　　　　二级景区1处 停留时间共40分钟（每处景区40分钟）
　　　　　休息站及景点4处 停留时间40分钟（每处景区10分钟）

四级游览路线：40分钟—1小时
　乘坐游船时间：40分钟—1小时

● 一级游览景区　● 二级游览景区　● 景点及休息服务站

South entrance scenic route analysis:
Primary scenic route: 4-5h, 5-6km (motor vehicles managed by the Park available)
— Route A: 6km (you can take a yacht to return on the way)
　　Bicycle: 30 minutes (3h 50min for visiting all the scenic spots)
　　Battery car: 1h (4h 20min for visiting all the scenic spots)
　　Walk: 2h (5h 20min for visiting all the scenic spots)
　　Pass: 3 primary scenic spots, 2h of stay (40min for each scenic spot)
　　　　2 secondary scenic spots, 40min of stay (20min for each scenic spot)
　　　　4 comfort stations and scenic spots, 50min of stay (10min for each scenic spot)
--- Route B: 6km (you can take a yacht to return on the way)
　　Bicycle: 30 minutes (3h 10min for visiting all the scenic spots)
　　Battery car: 1h (3h 30min for visiting all the scenic spots)
　　Walk: 2h (4h 30min for visiting all the scenic spots)
　　Pass: 2 primary scenic spots, 1h 20min of stay (40min for each scenic spot)
　　　　2 secondary scenic spots, 40min of stay (20min for each scenic spot)
　　　　3 comfort stations and scenic spots, 30min of stay (10min for each scenic spot)
--- Route C: 6km (you can take a yacht to return on the way)
　　Bicycle: 30 minutes (3h 20min for visiting all the scenic spots)
　　Battery car: 1h (3h 50min for visiting all the scenic spots)
　　Walk: 2h (4h 50min for visiting all the scenic spots)
　　Pass: 2 primary scenic spots, 1h 20min of stay (40min for each scenic spot)
　　　　3 secondary scenic spots, 1h of stay (20min for each scenic spot)
　　　　3 comfort stations and scenic spots, 30min of stay (10min for each scenic spot)
Secondary scenic route: 3-4h, 3km
— Route A: 3km (you can take a yacht to return on the way)
　　Bicycle: 15 minutes (2h 55min for visiting all the scenic spots)
　　Battery car: 30min (3h 10min for visiting all the scenic spots)
　　Walk: 1h (3h 40min for visiting all the scenic spots)
　　Pass: 2 primary scenic spots, 1h 20min of stay (40min for each scenic spot)
　　　　2 secondary scenic spots, 40min of stay (20min for each scenic spot)
　　　　4 comfort stations and scenic spots, 40min of stay (10min for each scenic spot)
Tertiary scenic route: 2-3h, 2km
--- Route A: 2km (you can take a yacht to return on the way)
　　Walk: 40min (2h 20min for visiting all the scenic spots)
　　Pass: 1 primary scenic spots, 40min of stay (40min for each scenic spot)
　　　　1 secondary scenic spots, 20min of stay (20min for each scenic spot)
　　　　4 comfort stations and scenic spots, 40min of stay (10min for each scenic spot)
— Route B: 2km (you can take a yacht to return on the way)
　　Walk: 40min (2h 20min for visiting all the scenic spots)
　　Pass: 1 secondary scenic spots, 20min of stay (20min for each scenic spot)
　　　　4 comfort stations and scenic spots, 40min of stay (10min for each scenic spot)
Quaternary scenic route: 40min-1h
　　Boat tour: 40min-1h

● primary scenic spots　● secondary scenic spots　● Scenic spot and comfort station

效果图 Effect Drawing

Forest park

奥林匹克森林公园湿地系统详图

wetland plan

	生物水处理设施 Biological Water Treatment Facility
	山脊（山顶） Ridge Top With Trail
	滩地和溪流 Riparcan Area(1)
	滩地和溪流 Riparian Area(2)
	沉沙池 De-silting Basin
	南坡林地 South Slope Forest
	功能性湿地 Treatment Wetland
	循环功能性湿地 Recirculating Treatment Wetland
	北坡林地 North Slope Forest
	三角洲湿地 Wetlands In Delta Area of Lake

平面图 Plan

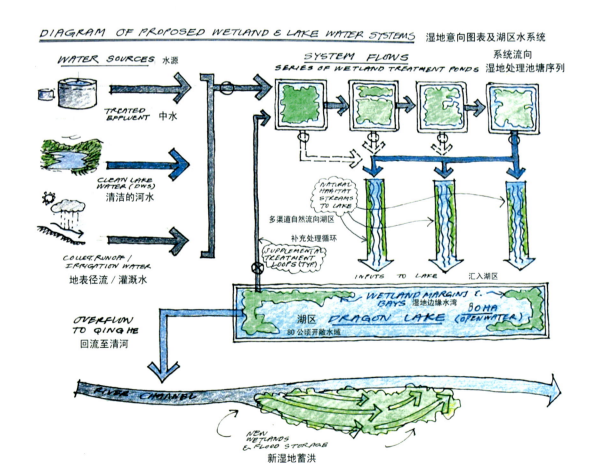

奥林匹克森林公园湿地系统详图
Forest Park Wetland Plan

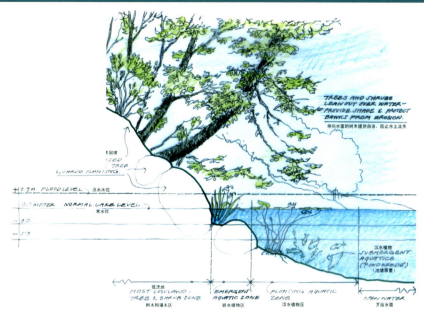

剖面图一
Profile Chart 1

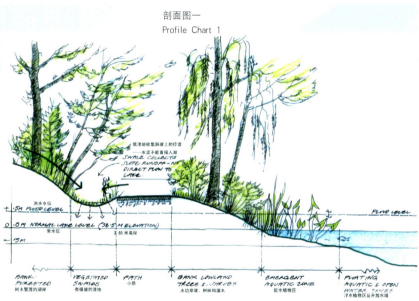

剖面图二
Profile Chart 2

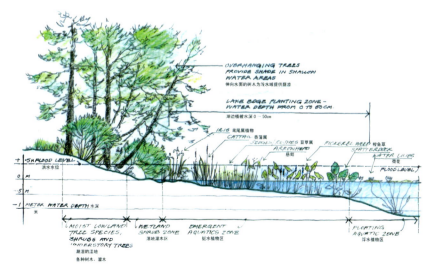

剖面图三
Profile Chart 3

奥林匹克森林公园"卧龙谷"[奥运纪念园]及宿根花卉区详图
Detail Plan the "Crouching Dragon Valley" (Olympic memory park)

奥林匹克森林公园龙涎涧及九江汇翠节点详图
Detail Plan of Longxianjian and Jiujianghuicui Gardens—forest Park

龙涎涧景观效果图
Perspective of Longxianjian

龙王堂及石碑景观效果图
Water Dragon Temple and Menument Perspective

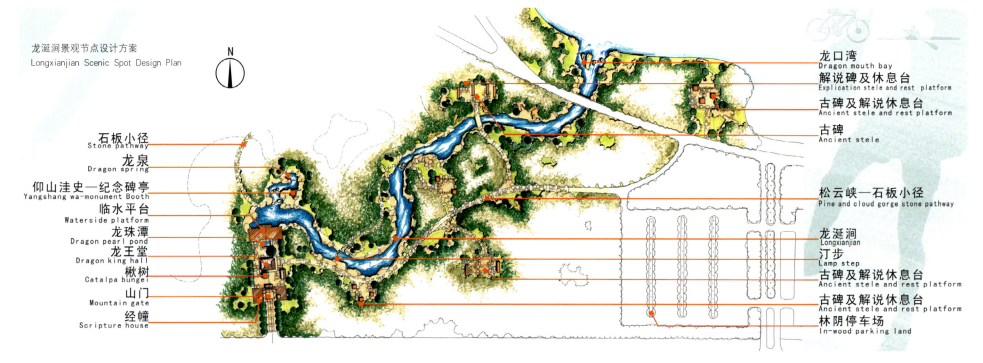

龙涎涧景观节点设计方案
Longxianjian Scenic Spot Design Plan

石板小径 Stone pathway
龙泉 Dragon spring
仰山洼史—纪念碑亭 Yangshang wa-monument Booth
临水平台 Waterside platform
龙珠潭 Dragon pearl pond
龙王堂 Dragon king hall
楸树 Catalpa bungei
山门 Mountain gate
经幢 Scripture house

龙口湾 Dragon mouth bay
解说碑及休息台 Explication stele and rest platform
古碑及解说休息台 Ancient stele and rest platform
古碑 Ancient stele
松云峡—石板小径 Pine and cloud gorge stone pathway
龙涎涧 Longxianjian
汀步 Lamp step
古碑及解说休息台 Ancient stele and rest platform
古碑及解说休息台 Ancient stele and rest platform
林阴停车场 In-wood parking land

九江汇翠景观节点设计方案
Jiujiang Scenic Spot Design Plan

九江汇翠景观效果图
Perspective of Jiujianghuicui

景观塔设计方案一
View Tower Plan One

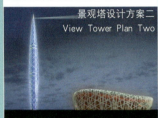
景观塔设计方案二
View Tower Plan Two

景观塔设计方案三
View Tower Plan Three

Plaza

奥林匹克森林公园中心区"舞"之广场,"火"之广场及观景塔详图
Detail Plan of "Dance Plaza","Fire Plaza" and the Observation Tower in Central Zone

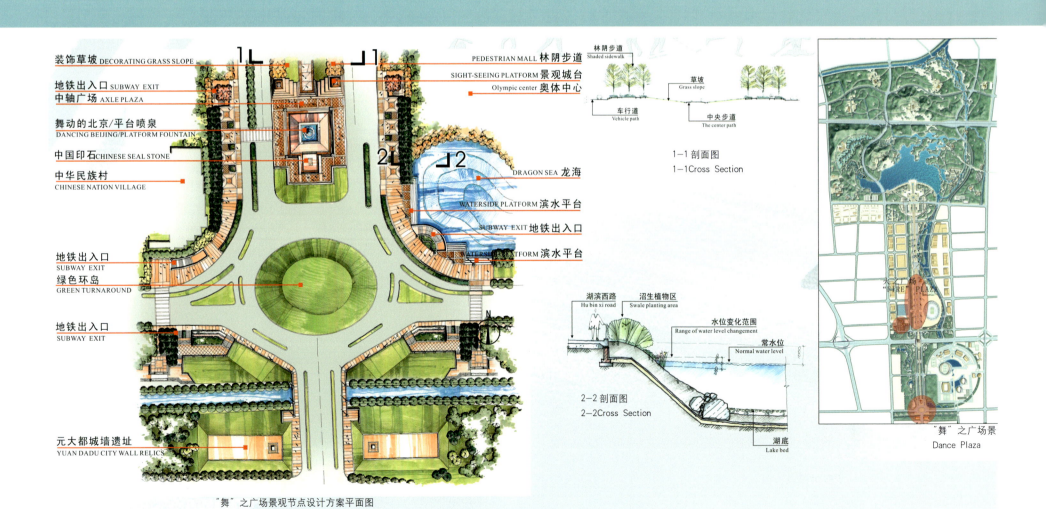

"舞"之广场景观节点设计方案平面图
Dance Plaza Scenic Spot Design Plan

"舞"之广场景
Dance Plaza

狂欢草坪
JAMBOREE GRASSLAND

青春北顶/北顶娘娘庙
YOUTH BEIDING / BEIJING NIANGNIANG TEMPLE

民族体育园
FOLK SPORTS PARK

庙前广场 TEMPER PLAZA

四环路绿化带
4TH RING ROAD GREEN BELT

娘娘庙，民族体育园平面图
Niangniang Temple, Folk Gymnasium Plan

北顶娘娘庙——先民举行盛会的地方
Beiding Liangliang Temple: The place where ancient people held the grand assemblies.

杂耍，高跷，龙灯是庙会业主要的活动。时代变迁 在先民盛会的地方，更大规模的盛会——奥运会开始了
Vaudeville, walking on stilts, and dragon lantern dance are the main events in the temple fair. With the change of the time, the grand assembly of larger-scale Olympic Game will be held in the same place.

奥运会——新世纪的国际盛会。青春女神——碧霞元君呼唤生命的激情与活力，为生命青春祝福。
The Olympic Game □ The international gathering in the new century. Goddess of Youth □ Bixiayuan calls the enthusiasm and vigor of life and blesses the life and the youth.

娘娘庙前绿地辟为民族体育体育园，以民族体育运动为主题，将原庙前戏台延伸为比赛表演场地。
National Sports Park will be held on the green land in front of Liangliang Temple and the former stage there will be changed into the place for performance and games.

原立于北顶庙中的《重修北顶娘娘庙记》、《永安老会碑》、《攒香碑记》记述了每年阴历四月十五日开始的庙会盛况，庙会持续5天，其中有自大屯树观音寺至北顶庙的游行。
The inscriptions of "To note the renovation of Beiding Liangliang Temple", "Inscription of Yong an Fair", and "Inscription of Zhanxiang Fair" formerly stood in Beiding Temple recorded the spectacular events of the temple fairs begun in every April 15 of the lunar year. The temple hair lasted 5 days and the events included the parade from Avalokitesvara Temple in Datunshu to Beiding Temple.

娘娘庙现状照片
The Current Picture of Niangniang Temple

景观塔设计方案一
View Tower Plan One

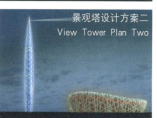
景观塔设计方案二
View Tower Plan Two

景观塔设计方案三
View Tower Plan Three

娘娘庙改造效果图
The Renovation Perspective Niangniang of Temple

"火"之广场景观节点设计方案
Fire Plaza Scenic Spot Design Plan

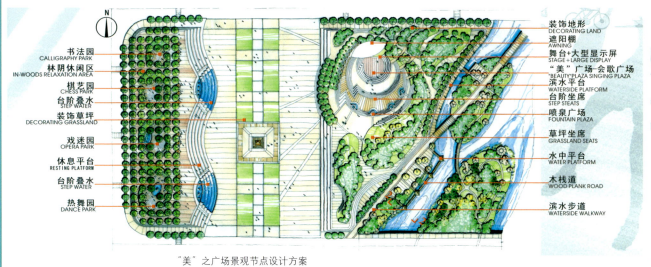

Central Zone of the Olympic Green

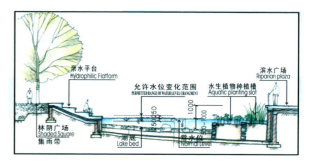

1-1 剖面图
1-1 Cross Section

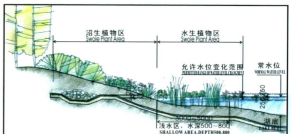

2-2 剖面图
2-2 Cross Section

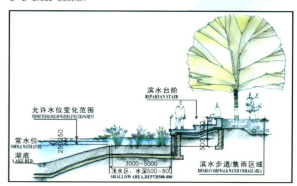

湖滨东路剖面示意图
Cross Section of HU BIN Eastern Road

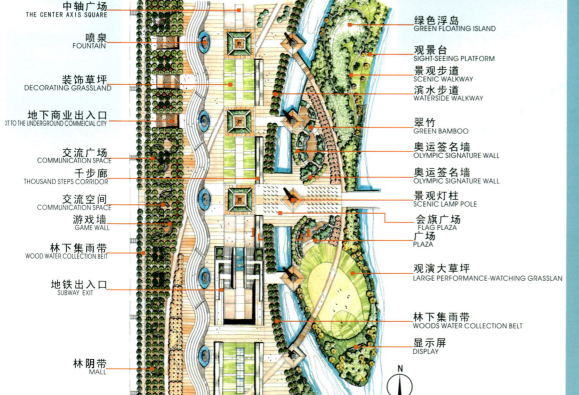

奥林匹克森林公园"美"之广场和"爱"之广场节点详图
Detailed Plan of Beauty Plaza and Love Plaza Nodes

奥林匹克森林公园及中心区全园鸟瞰夜景
Birds-eye Night View of Forest Park and Central Zone

奥林匹克森林公园中心区"绿"之广场景区设计方案
"Green plaza" Scenic Area Plan

A07

Beijing Institute of Landscape and Traditional Architectural Design and Research

北京市园林古建设计研究院

URS Australia Pty Ltd.

URS 澳大利亚有限责任公司

奥林匹克公园森林公园及中心区景观规划设计方案

OLYMPIC GREEN LANDSCAPE DESIGN FOREST PARK AND CENTRAL ZONE

Dragon—Phoenix

龙凤概念图
Dragon—Phoenix

1 史之门 HISTORY GATEWAY
2 历史区 HISTORIC SECTOR
3 四环路标志 FOURTH RING ROAD MARKER
4 北顶娘娘庙及天寺庙园林 BEIDING GOLDEN TEMPLE AND PARK
5 果林 THE ORCHARD
6 国家体育场 NATIONAL STADIUM PRECINCT
7 龙形水系-龙腹南部 BELLY OF THE DRAGON (SOUTHERN SECTOR)
8 体育区 SPORTS SECTOR
9 中一路标志 ZHONGYI ROAD MARKER
10 龙形水系-龙腹北部 BELLY OF THE DRAGON (NORTHERN SECTOR)
11 文化区 CULTURAL SECTOR
12 北一路标志 BEIYI ROAD MARKER
13 露天阶梯看台 TERRACES
14 湖滨广场 LAKE SECTOR
15 开放公园 URBAN PARK
16 龙头湖南岸 DRAGON LAKE SOUTHERN SIDE PRECINCT
17 龙王庙 TEMPLE OF THE DRAGON KING
18 湿地 WETLANDS
19 龙头湖面(西部) DRAGON LAKE (WEST)
20 生态教育中心 ENVIRONMENTAL CENTRE
21 凤尾 PHOENIX TAIL
22 跨五环绿色走廊(南) LAND BRIDGE
23 具身与凤翅 PHOENIX BODY AND WING
24 青栩落叶混交林 PINE OAK FOREST
25 东北部小山 NORTH-EAST HILL
26 五环路以北水面 NORTHERN PONDS
27 "凤眼"标志 EYE OF THE PHOENIX MARKER
28 森林公园跨五环通道 FOREST GATEWAY
29 龙角水面 NORTHERN LAKE PONDS
30 "龙眼"标志 EYE OF THE DRAGON MARKER
31 龙头湖面(东部) DRAGON LAKE (EAST)
32 龙头湖面东岸 EASTERN SHORE OF THE DRAGON LAKE
33 玉苑别墅 GITU PARK VILLA
34 清河等线明渠带状公园 QINGHE LINEAR PARK
35 清河河道 QINGHE RIVER CORRIDOR

DESIGN CONCEPT
设计理念

"绿色奥运"将通过运用生态、社会和经济可持续平衡发展原则来体现，使奥林匹克公园成为其与周边地区可持续性发展的重要推动力，我们的设计将集中在这些领域树立榜样：

* 市政规划和设计。
* 交通规划。
* 景观设计与管理。
* 建筑物设计。
* 节能和使用可再生能源技术。
* 材料选用。
* 水资源管理和中水回用。
* 固体废弃物管理和资源回收。

"科技奥运"将通过运用一系列适合当地情况的新技术的应用来体现。这不但包括高科技的应用，在合适情况还包括成熟而价格适中，具有可持续运营特色的中等技术的应用。

技术的应用包括在以下领域：

* 废水回收利用。
* 可再生能源技术，如太阳能、沼气和风能的利用。
* 低温室气体排放能源，如天然气使用。
* 在冬季湿地对中水清洁过滤功能下降时运用纤维过滤技术清洁过滤中水。
* 沙石和水生植物过滤技术，用于清洁中水和雨洪。
* 不造成污染和高效能的交通系统。
* 最先进的通信技术（视听技术），使公园游客能在园区和北京市各处实况观看奥运比赛。
* 能为游客提供各种不同有关中国和北京体育、文化和历史性信息的通信技术。
* 向游客介绍奥林匹克公园、北京和中国各地有关环境保护信息的通信技术。
* 处理和回收固体废料的技术。
* 结构土壤技术，使在需要硬质铺装的地面种植的树苗能健康成长，同时不对铺装产生影响（如拱裂地面）。
* 高科技的照明技术，方便游客在晚间前来奥林匹克公园游览并为之带来独特的体验。
* 节水型高科技灌溉技术，用于灌溉大面积的景观园林区。
* 高科技的水上特色景观和水质管理系统,减少水质管理过程中的耗能和降低化学添加剂的使用。

"人文奥运"将通过各种设施的建立和管理计划为游人创造一个愉快、舒适安全的观赛和游览环境，并使游人能够参加各种文化、娱乐和学习活动。人文奥运会的成功与否很大程度上取决于中外游客能否方便自由的交往。本设计充分考虑到人工和天然公共绿地的规划设计能满足当地和海外家庭、年轻人、老年人和残疾人的游览娱乐需要。

人文奥运还将反映中国文化的多样性和海外游客所代表的世界各地丰富多彩的文化。本设计将兼顾到下列各方面的考虑：

* 充分显示和介绍历史上和现代中华文明的成就。
* 在中心区通过条形树码的种植和解说演示系统介绍中国55个少数民族的风俗习惯和他们对中华文明的贡献。
* 为游客和家庭在奥运会期间和赛后提供各种设施。
* 为儿童建设各类游乐设施和互动型解说设施。
* 为年轻人建设聚会和社交设施。
* 在整个园区安装标识和信息介绍系统，方便游人辨认方向和了解公园建设的各种含义。
* 为晚间游览提供方便。
* 为周边居民前来园区游览散步和进行自行车锻炼建设路网。

森林公园鸟瞰图
Forest Park Bird's eye Views

'The Green Olympics' will be achieved by incorporating the principles of ecological, social and economic sustainability throughout the whole site and through its connections to surrounding areas of Beijing. This will be achieved through excellence in:

* Urban planning and design.
* Transportation planning.
* Landscape design and management.
* Building design.
* Energy conservation and renewable sources.
* Materials selection.
* Water management and reuse.
* Solid waste management and resource recycling.

'The High-Tech Olympics' will be achieved by applying technology that is appropriate to the situation. This will not only involve high level technology but also in some appropriate situations it will involve mid-level technologies that are cost-effective, sustainable and well tested.

The application of Technology will include:
* Waste water reuse/recycling.
* Renewable energy sources including solar, geothermal, biogas and wind.
* Low greenhouse gas emitting energy such as natural gas.
* Membrane technology for waste water treatment during winter months when wetlands are not functioning.
* Gravel and reed bed technology to polish wastewater/stormwater.
* Transport technology that is non-polluting and energy efficient.
* Communications technology (audio, visual) that is leading edge to allow visitors to watch Olympic events throughout the Olympic Green site and other live sites throughout Beijing.
* Communications technology to provide a rich mixture of information about the sporting, cultural and historic aspects of Beijing and China.
* Communications technology that will help visitors to learn about the environment of the Olympic Green, Beijing and other parts of China.
* Technology to treat and recycle solid waste.
* Structural soil technology to allow trees to grow well in paved areas without causing damage to the pavement.
* High technology lighting to enhance the night time experience of visitors to the Olympic Green.
* High tech irrigation systems to minimise the use of water for irrigation of landscaped areas.
* High technology water features that minimise the requirement for energy to pump water and chemicals to treat the water.

The People's Olympics' will be achieved by creating the physical facilities together with operational programs that allow enjoyable, safe, convenient participation in the sporting events as well as the cultural and recreational and learning activities. The success of the Peoples' Olympics will be demonstrated by high levels of interaction between visitors from Beijing, China and the rest of the world. Planning and design of the urban and natural environments will provide for the needs of all categories of visitor including local and foreign families, youth, elderly people, and those with handicaps.

The cultural diversity of the Chinese people, as well as visitors from all parts of the world is to be acknowledged and welcomed at the People's Olympics. This will involve:

* Rich layers of information and interpretation of the achievement of the Chinese people throughout history and in modern times.
* The 55 Ethnic Groups of China will be celebrated by distinctive strips of trees across the Central Zone together with presentation of interpretation information at each of the tree strips.
* Facilities for families visiting the Olympic Green both during and after the Olympic Games.
* Facilities for children including playgrounds and interactive interpretation facilities.

Sites and facilities for youths to meet and socialise.

Signage and information to allow people to find their way around the site conveniently and safely.

Design for night time use of the Olympic Green site.

Convenient access to the site from surrounding residential areas by walking, cycling and public transport.

概念：龙与凤

本设计方案将现代景观设计手段与传统中国景观的基本元素结合起来。它反映的是景观设计可以在吸取中华文化五千年文明之精华的同时，能成功运用现代设计理念处理和解决环保问题。

中国社会对环保问题的重视将充分反映在本设计所制定的恢复和重建奥林匹克公园园区生态系统的各种可持续运营措施和新技术的应用上面。同时，我们的设计无论在使用自然元素或文化元素来创造景观上，都将独树一帜。

在运用传统文化元素方面，本设计在公园原有龙形水系总体规划基础上加上了凤凰山这一景观造型，使中华文明里的两大文化象征得到平衡和利用。运用现代手段通过堆造凤凰山，植树造林和创建龙形水系来创造景观空间是符合传统中国的园林设计宗旨的。我们的设计还将广泛使用诸如太阳能技术和水资源循环利用等可持续运营手段来为奥林匹克公园增色。

龙与凤，这两个强烈的中华文化象征所代表的是阴和阳。龙是负责治水的，而凤则代表恢复生机和修复大地。奥林匹克公园总体规划原始方案征集里胜出的龙形水系设计现通过本方案在北部建设凤凰山而得到平衡和完善。

具有强烈艺术风格的凤翅的建造原料将来自于开挖龙湖所需要处理的土方。凤身宛然穿越五环路的大型旱桥，使森林公园的北区和龙湖西岸的森林公园西区连为一体，野生动物能够自由在两区之间穿行，游人也能够方便地来往于两个游览区。凤尾则围住公园的湿地，连接公园的南北两个游览区。

为充分显示凤凰造型，本设计将在凤凰山四周种植森林，而凤凰山体的处理，则是在北坡种植低矮草本植物，因为这类植物较善于保持土壤内的水份。山体南坡可种植较高的禾本科植物，这类植物也能适应相对干旱的生长条件。

对位于中轴线上成为本设计中五个标志性结构之一龙眼的处理，我们采用圆形。作为中轴线上最北的标志性结构的凤眼，我们也将之变为五环以北的一个特色景观区。

Phoenix

The Dragon and the Phoenix

This concept combines contemporary landscape design with the fundamental elements of traditional Chinese landscape design. It reflects the increasingly successful application of contemporary design ideas while drawing on the cultural legacy of 5,000 years of Chinese cultural development.

China's changing attitude to the environment is reflected in ecological restoration of the site and adoption of sustainability principles both in the natural and cultural landscape

Traditional values are reflected through incorporation of the dragon of the masterplan and balancing it with a symbol of the phoenix in the northern landscape. The creation of space using planting, earth mounding and water applies a traditional Chinese landscape design approach in a contemporary manner. The application of technology such as solar energy and closed cycle waste recycling systems.

The dragon and the phoenix , these strong symbols reflect Ying and Yang. The Dragon responsible for water floods and droughts. The Phoenix is a symbol of recovery and restoration. The Dragon which forms an integral part of the winning masterplan has been balanced by the Phoenix to be created in the northern portion of the site.

The styalised shape of the wings of the Phoenix are formed by earth mounds created largely from material excavated to construct the lake. The body of the Phoenix extends across the Fifth Ring Road by a land bridge that will allow wildlife movement as well as people using the Park to become integrated with the western edge of the lake. The tail wraps around the wetlands to integrate the southern and northern portions of the Forest Park.

The form of the phoenix is defined by forest vegetation surrounding the earth mounds. The earth mounds are to be planted with low meadow plants generally on the north-facing slopes which generally retain more soil moisture with the south facing slopes planted predominantly with grasses to take account of the relatively dry growing conditions.

The eye of the dragon, which is located on the central axis is formed by one of the five markers and is circular in shape. The eye of the phoenix is also formed by the northern most marker, which is also located on the central axis, but north of the Fifth Ring Road.

Dragon-Phoenix

总体规划图
Overall Planning Chart

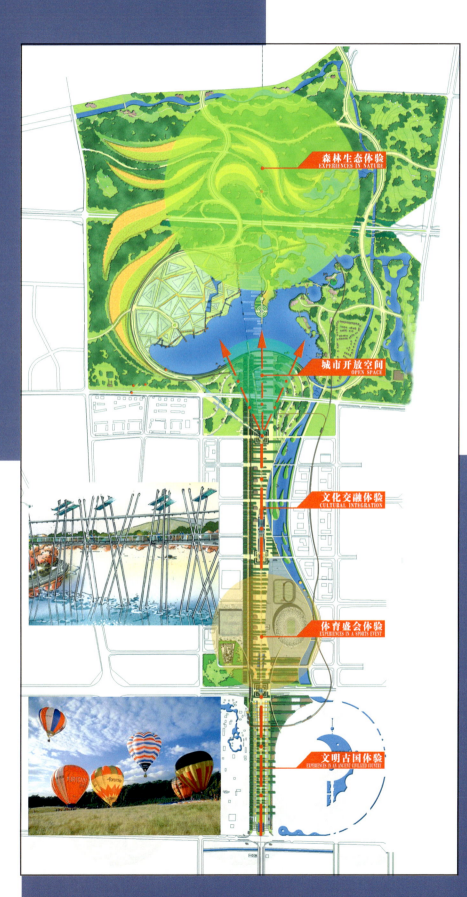

景观规划分析图
Landscape Planning Assay Map

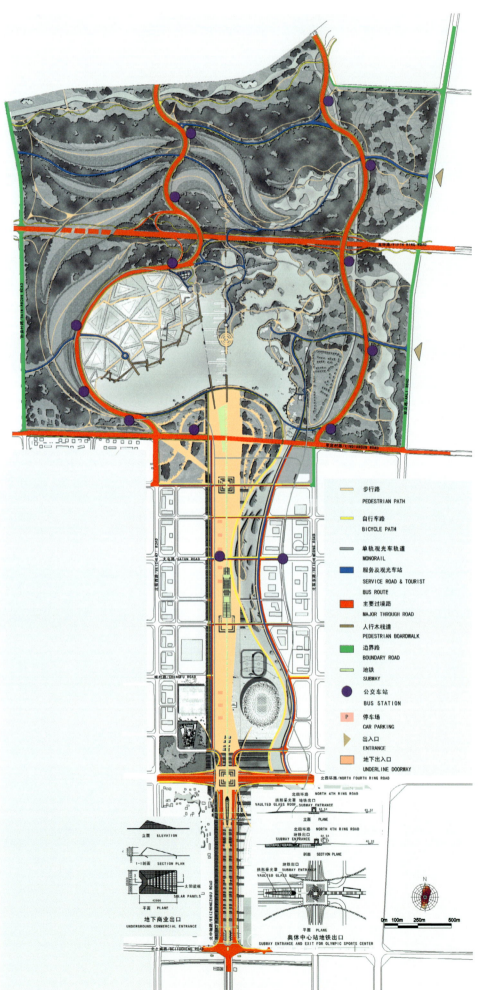

交通系统及交通设施规划图
Traffic Planning Chart

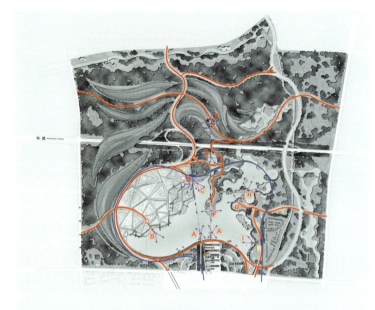

浏览路线分析图
Tourist Routes and Scenic Spots

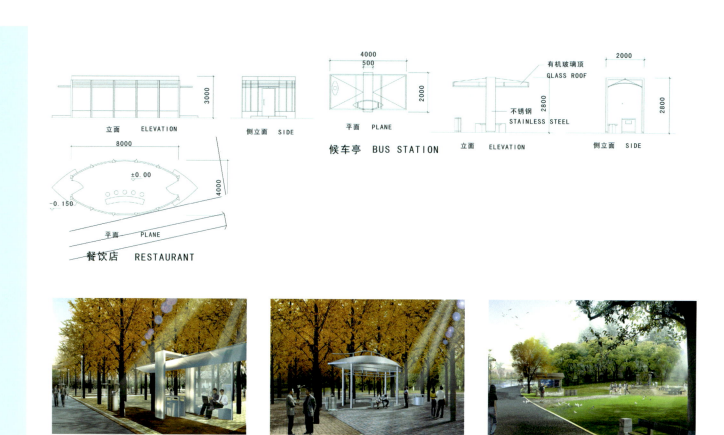

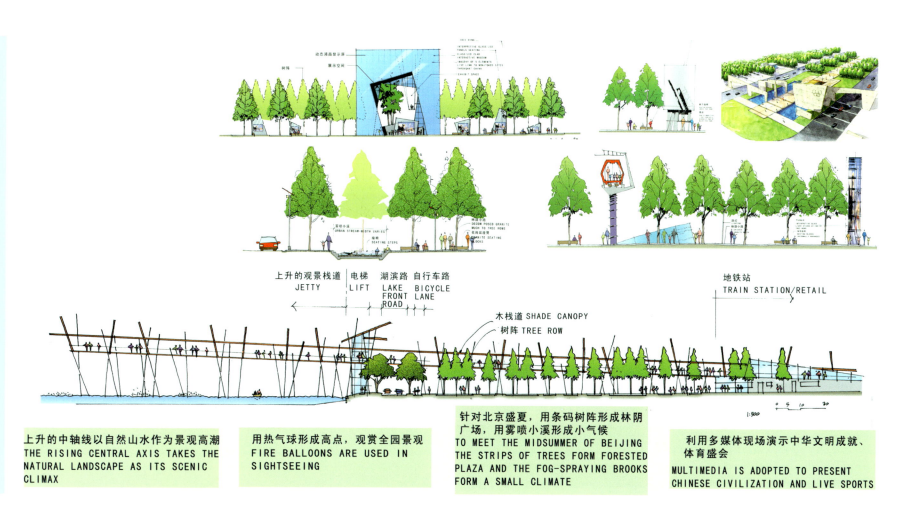

| 上升的中轴线以自然山水作为景观高潮 THE RISING CENTRAL AXIS TAKES THE NATURAL LANDSCAPE AS ITS SCENIC CLIMAX | 用热气球形成高点，观赏全园景观 FIRE BALLOONS ARE USED IN SIGHTSEEING | 针对北京盛夏，用条码树阵形成林阴广场，用雾喷小溪形成小气候 TO MEET THE MIDSUMMER OF BEIJING THE STRIPS OF TREES FORM FORESTED PLAZA AND THE FOG-SPRAYING BROOKS FORM A SMALL CLIMATE | 利用多媒体现场演示中华文明成就、体育盛会 MULTIMEDIA IS ADOPTED TO PRESENT CHINESE CIVILIZATION AND LIVE SPORTS |

森林公园夜景鸟瞰图
Forest Park Nightscape Lighting

规划总平面图——中心区
Overall Planning Chart – Central Zone

规划总平面图——森林公园
Overall Planning Chart - Forest Park

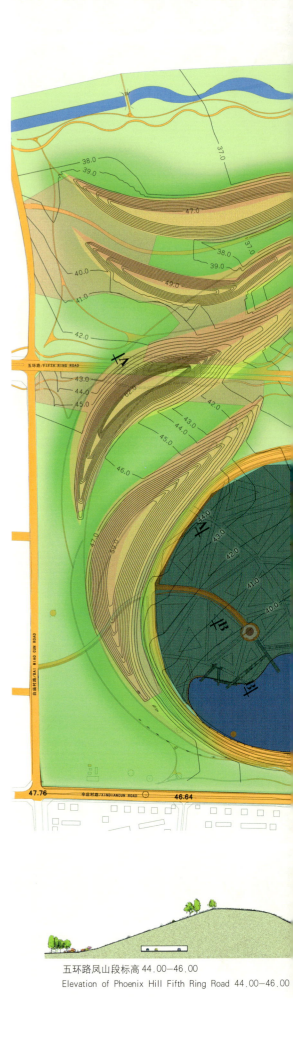

五环路凤山段标高 44.00—46.00
Elevation of Phoenix Hill Fifth Ring Road 44.00—46.00

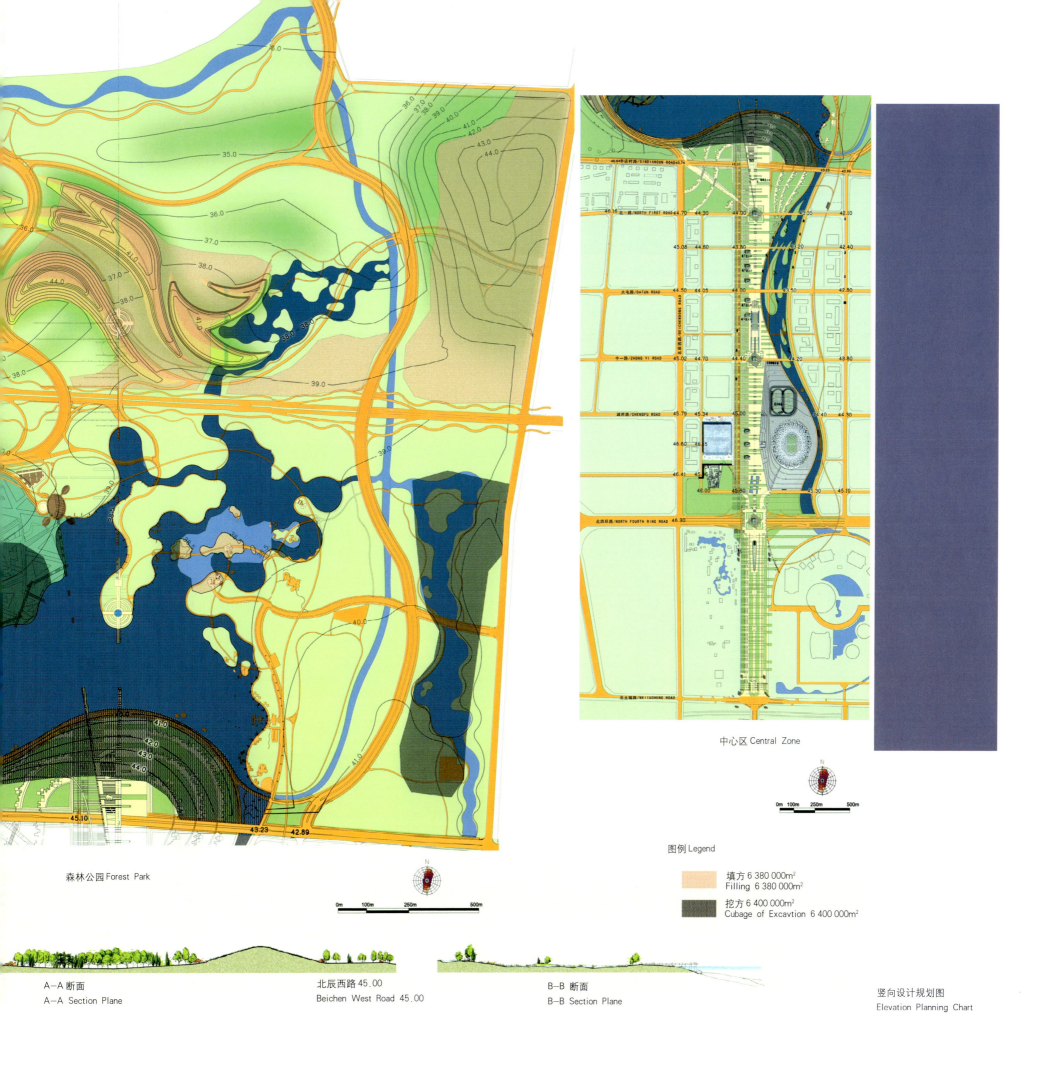

森林公园 Forest Park

中心区 Central Zone

图例 Legend

填方 6 380 000m²
Filling 6 380 000m²

挖方 6 400 000m²
Cubage of Excavtion 6 400 000m²

A-A 断面
A-A Section Plane

北辰西路 45.00
Beichen West Road 45.00

B-B 断面
B-B Section Plane

竖向设计规划图
Elevation Planning Chart

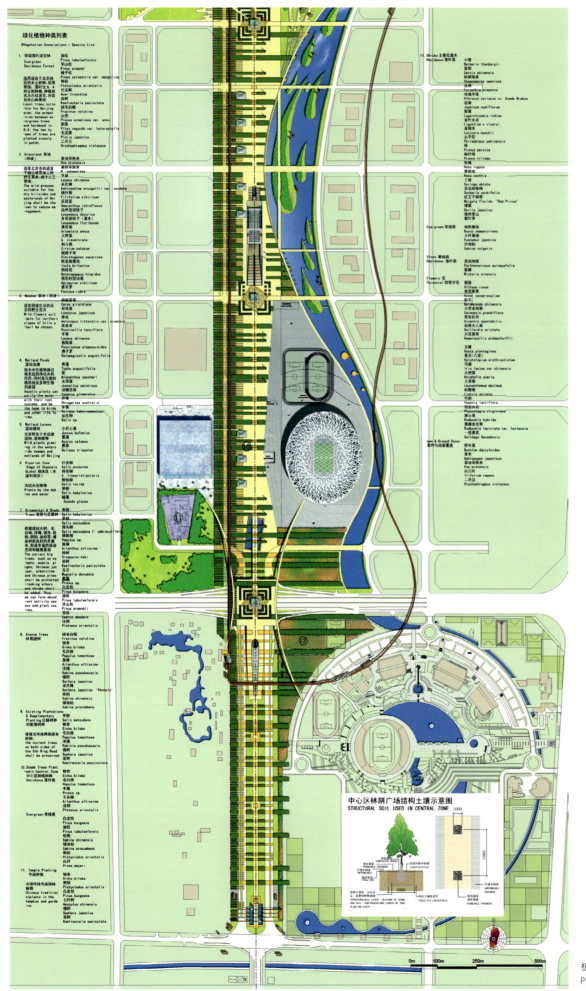

植物种植设计规划图
Plant Planning Chart

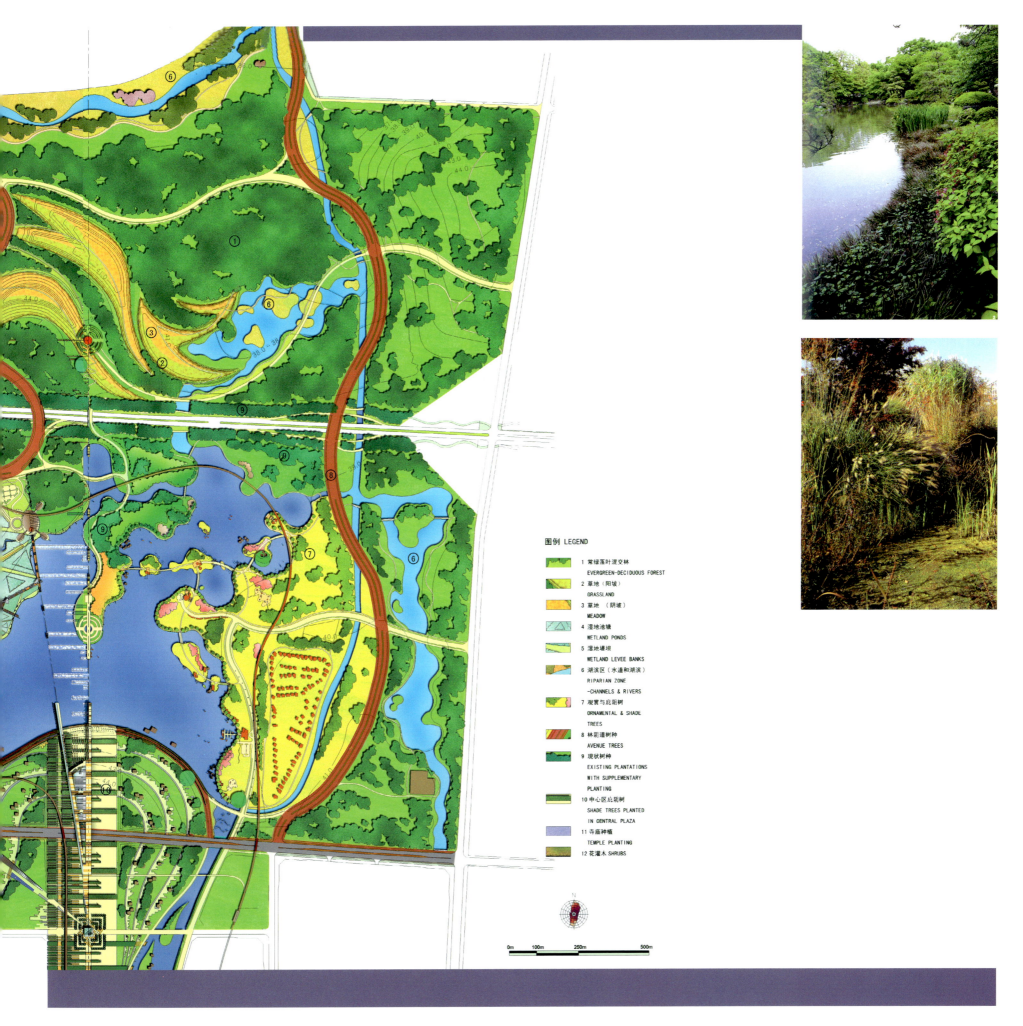

森林公园鸟瞰图
Foerst Park Bird's Eye View

WATER MANAGEMENT | 水管理

LEGEND | 图例

	1 WATER CONTROL GATE	水闸门控制
	2 WETLAND AREA	湿地区
	3 FLOW DIRECTION THROUGH WETLANDS	穿过湿地的水流方向
	4 OPEN CHANNLE	明渠
	5 PIPELINE	水管
	6 SURFACE RUNOFF DIRECTION	表面径流方向
	7 RECYCLED WASTE WATER FROM SEWAGE TREATMENT PLANT	中水
	8 STORMWATER DETENTION BASINS	雨水汇集池
	9 WATER SUPPLY FROM RIVER	从河流中供水的设施
	10 PUMPING STATION (SOLAR POWERED)	太阳能供电的泵站
	11 PIPES BELOW FIFTH RING ROAD	五环路下面的管道
	12 DISCHARGE POINT	排放点
	13 SUPPLY FROM SEWAGE TREATMENT PLANT	从污水处理厂供水点
	14 RIVER	河流
	15 DRAINAGE CHANNEL	排水渠
	16 ROADWAY WITH PERMEABLE PAVING	可以渗透雨水的道路
	17 PARKING AREA WITH PERMEABLE PAVING	可以渗透雨水的停车场

LAKE EDGE TREATMENTS | 湖边处理

	18 NATURAL EDGE FORMED BY LEVEE BANK & ROCK PROTECTION	用防洪堤岸或者岩石护堤围成的自然驳岸
	19 NATURAL EDGE ROCK & PLANTING ON BANK	用岩石和绿化树木筑成的自然驳岸
	20 MIXTURE OF SOFT EDGE WITH ROCK & PLANTING ALTERNATING WITH FORMAL TREATMENT OF STONE	用岩石和绿化树木筑成的自然驳岸，间或错落有序铺设石阶
	21 FORMAL EDGE WITH STONE PAVING AND STEPPED TERRACES ALONG LAKE EDGE	用石头铺设的硬质驳岸，间或沿湖边筑成的台地

Dragon-Phoenix

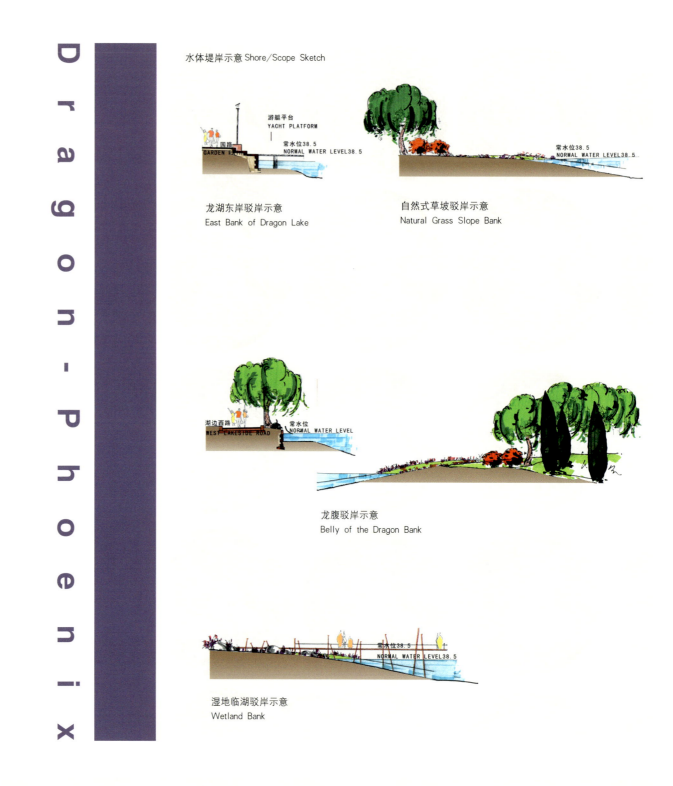

水体堤岸示意 Shore/Scope Sketch

龙湖东岸驳岸示意
East Bank of Dragon Lake

自然式草坡驳岸示意
Natural Grass Slope Bank

龙腹驳岸示意
Belly of the Dragon Bank

湿地临湖驳岸示意
Wetland Bank

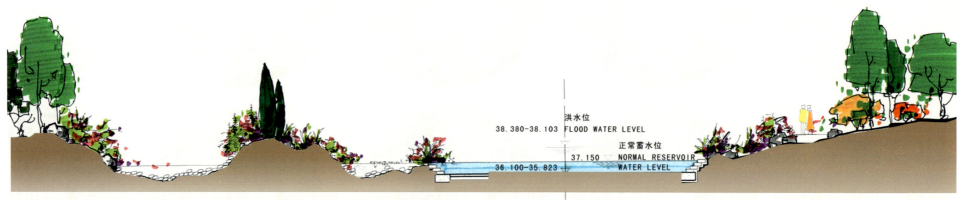

清河导流渠驳岸示意
Qinghe Diversion Channel Bank

赛时规划总平面图
General Layout during the Olympic Games

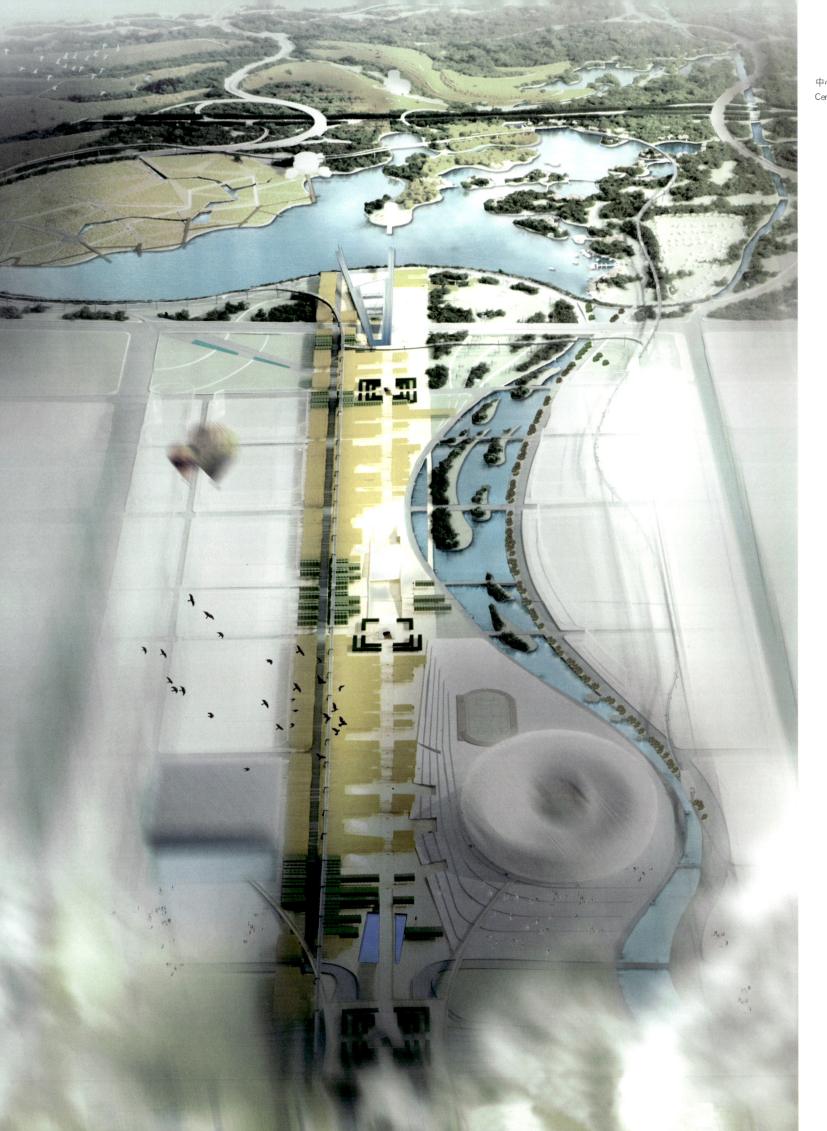

中心区鸟瞰图
Central Bird's Eye Views

中心区夜景鸟瞰图
Central nightscape lighting

Dragon–Phoenix

后 记

  此次《北京奥林匹克公园森林公园及中心区景观规划设计方案征集》活动，得到国内外知名景观设计单位的热情支持和参与。2003年7月份开始征集工作，2003年11月21日完成方案评审和公开展览。此后，经有关部门研究确定以美国Sasaki公司和清华城市规划设计研究院合作方案为基础，借鉴综合其他方案的特点和优点完善调整深化作为实施方案。此次征集过程，凝聚着众多参与此项工作的单位、个人的辛勤劳动和智慧。在此，我们再次向他们表示衷心的感谢！

编 者
2003年12月30日

POSTSCRIPT

  Our soliciting landscape planning and design schemes of the Forest Park and the Central Zone in the Olympic Green includes the enthusiastic support and active participation of well-known landscape design agencies both at home and abroad. The process started in July 2003, with design schemes assessed and exhibited to the public on November 21, 2003. As reviewed by responsible departments, it was later on recommended that the scheme submitted by Sasaki Associates,Inc. and Beijing Tsinghua Planning Cororation, be tuned and perfected, as a scheme for action, by incorporating the characteristics and advantages of other schemes. This process included the diligence and wisdom of the organizations and individuals involved. We wish to express our heartfelt gratitude to them again!

Editor
December 30, 2003